Progress in Optical Science and Photonics

Volume 11

The purpose of the series Progress in Optical Science and Photonics is to provide a forum to disseminate the latest research findings in various areas of Optics and its applications. The intended audience are physicists, electrical and electronic engineers, applied mathematicians, biomedical engineers, and advanced graduate students.

More information about this series at http://www.springer.com/series/10091

Kamal Nain Chopra

Optoelectronic Gyroscopes

Design and Applications

Springer

Kamal Nain Chopra
Laser Science and Technology Centre
New Delhi, India

ISSN 2363-5096 ISSN 2363-510X (electronic)
Progress in Optical Science and Photonics
ISBN 978-981-15-8379-7 ISBN 978-981-15-8380-3 (eBook)
https://doi.org/10.1007/978-981-15-8380-3

This Springer imprint is published by the registered company Springer Nature Singapore Pte Ltd.
The registered company address is: 152 Beach Road, #21-01/04 Gateway East, Singapore 189721, Singapore

Foreword

It is a matter of great pleasure for me to express that Dr. Kamal Nain Chopra has made a tremendous and commendable effort in writing this book on optoelectronic gyroscopes and the related technologies, a topic, on which the availability of literature, especially at one place, is very much required for the DRDO scientists in particular and the researchers and academicians in general. Practically, all the concepts, including the novel evolving ones, characterization, and applications of photonic crystals along with the devices based on them, have been discussed briefly in this book, thereby making it very useful indeed for the scientific community in both India and abroad. It is clear from the literature that prior to this, very few attempts seem to have been made in presenting the different aspects of the subject from the research point of view, at one place, and therefore, this effort is expected to bridge the gaps between various types of research papers and literature on the subject at different places. The book should especially be useful for the designers and engineers of the optoelectronic gyroscopes, as the design aspects for the efficiency optimization have been presented and discussed.

In addition, the book is expected to be of great interest and utility for the budding researchers and scientists in the field, since it provides a large number of theoretical and experimental results available in the literature, for them to have a clear understanding of the subject and also to choose the direction in which to move for carrying out research in this fascinating field. It is my sincere wish that this book serves the researchers in enhancing their inputs on the topic and also their interest in making more concentrated efforts in carrying out research in this rapidly evolving field.

Prof. Dr. Vipin Kumar Tripathi
Lasers and Plasmas Group,
Department of Physics, Indian
Institute of Technology, Delhi
New Delhi, India

Preface

A lot of literature, especially books on the concepts, technologies, and related phenomena of optoelectronic gyroscopes, is available. However, most of the books seem to have been written from the textbook point of view; not much effort seems to have been made on writing books for the researchers. Also, the premier books have been written a decade earlier or even before that. The literature for the studies conducted on the topic after that is quite scattered. This book is a sincere effort of the author to make available very useful information on this topic, especially from the research point of view, and also to bridge the gap between the information in the books and the research works done after that. A lot of emphasis has been laid on the research investigations carried out during the last decade, which makes this book very useful for the researchers to get an idea about the latest trends of research in this fascinating field. Also, the information has been presented in a brief and concise manner, making it easier for the researchers to understand the concepts quickly and go through the new papers after choosing the direction of interest.

In addition, the technical analysis of the theoretical aspects of concepts of the technologies used for making some of the components, especially ring laser mirrors, the substrates used and optical testing, and characterization of components has been presented.

In addition to the discussions regarding the theoretical modeling and designing of these devices, some related experimental results available in the literature have been presented to make the presentation clear and meaningful. All these devices are having many useful applications in various fields like sensors and industry, and in practical research applications in most of the scientific and engineering topics.

A number of important academic institutes like IITs, IISc, Bangalore, and universities, and scientific laboratories including National Physical Laboratory, other CSIR Laboratories, ISRO, and DRDO (in which LASTEC and IRDE have been quite active) are actively engaged in this subject. In view of such immense importance of the topic for so many institutes and scientific laboratories, a book on this topic is really needed, which will undoubtedly serve the purpose of understanding the complexities of this subject for the scientists of NPL, DRDO, and CSIR. It is also hoped that apart from being useful to these scientists and technologists, this book will serve as the motivating force for the researchers entering this complicated and rapidly evolving field. Also, the discussions on the theoretical modeling and designing of these devices, along with some related experimental results, should be quite useful for the designers and engineers engaged in developing devices based on sensors, optical testing, and thin-film characterization for the newer applications and more importantly research purposes.

Chapter 1 gives the optimization and mathematical modeling of ring laser gyroscopes (RLGs), for enhancing their efficiency. In addition, the problems encountered in the fabrication and working of the RLGs and some of the techniques for solving them have been highlighted. A brief qualitative review of the recent novel investigations on RLGs has also been given. This chapter is expected to be very useful for the new entrants in this fascinating field and also for the designers and technologists already engaged in improving the design of RLGs.

Chapter 2 gives the optimization and mathematical modeling of fiber optical gyroscopes (FOGs), along with results of some of the computations of the important parameters considered for designing the FOG. The problems encountered in the fabrication and working of the FOGs and some of the techniques for solving these problems have been highlighted. The sources of error in the FOG, some of the important results reported in the literature, and the designing aspects of the FOG have been technically discussed. Apart from presenting a qualitative review of the novel investigations on FOG, the recent innovations in finding the alternatives to the FOG have been discussed. The chapter is expected to be useful to the researchers and the designers in this fascinating field.

Chapter 3 gives the evolution of the laser coating technology and the development of the dielectric laser mirrors, which have drawn the attention of various researchers and also commercial firms. However, not much work seems to have been done on the development of the dielectric mirrors with very low scattering loss ($\sim$5–10 ppm). This chapter presents the experimental results, which have been observed and verified during the course of research and development for designing and fabrication of such coatings. A technical discussion of all the important aspects—optics, materials (coatings and substrates), designs, cleanliness conditions, coating techniques, and optimization of the process control parameters for the successful development of such coatings, with applications in the ring laser gyroscope used as the inertial navigation system—has also been given. The experimental results in improving the scattering loss by different techniques given in the chapter are on the basis of the long-term experience of the designing and fabricating the low scattering loss dielectric laser mirrors.

Chapter 4 gives the evolution of the high-power laser coating technology and the development of high laser damage threshold coatings, which have drawn the attention of various researchers. The present chapter discusses technically all the important aspects—optics, materials (coatings and substrates), designs, and coating techniques for the development of such coatings—and presents the important points, which have been observed and verified during the course of research and development for designing and fabrication of such coatings for various types of high-power lasers. The experimental results in improving the laser-induced damage threshold by different techniques have been presented in the paper.

Chapter 5 provides the technical analysis of the important techniques for optical testing of optical elements—Ronchi test, optical nondestructive testing (NDT), optical fiber NDT, laser speckle interferometry and speckle NDT, infrared thermography NDT, endoscopic NDT, terahertz (THz) NDT technology, improved interferometric optical testing, phase-shifting interferometry, phase-shifting single-shot interferometer technique, electronic speckle pattern interferometry (ESPI) technique, single-shot Fizeau interferometer technique, optical time-domain reflectometer, optical interferometry, and fiber-optic measurement technique. A brief discussion of the applications of optical testing has been given. A qualitative review of the recent novel studies on optical testing has also been included. The technical analysis and the overview should be of good utility to the new entrants in the field, and also the designers and engineers engaged in the design and development of high-quality optical elements and their testing.

Chapter 6 gives an overview of the important characterization techniques for optical thin films. Scientific analysis of various thin-film characterization techniques like X-ray photoelectron spectroscopy (XPS), secondary ion mass spectrometry (SIMS), scanning tunneling microscope, transmission electron microscope (TEM), reflection high-energy electron diffraction (RHEED), atomic force microscope (AFM), Fourier transform infrared (FTIR) technique, and differential interference contrast microscope has been given. Also, a brief discussion of the material characterization, structural features (macrostructure, mesostructure, microstructure, and nanostructure), and analytical characterization has been included. Some recent important studies on the scattering loss and the absorption loss of the optical thin films have also been briefly discussed. This overview should be of good utility to the new entrants in the development of high-quality optical thin films and their characterization.

New Delhi, India Kamal Nain Chopra

Acknowledgements

The author is grateful to DRDO in general and LASTEC in particular for providing an opportunity to work for many years with a number of scientists working on optoelectronic gyroscopes and the related technologies. Thanks are also due to the Photonics Group of the Indian Institute of Technology, Delhi, where the author got a very good exposure to the subject during the short period, working as Research Scientist in the group. A large number of presentations and discussions on the complexities and technicalities of some of the topics of optical gyroscopes and the related technologies have been immensely helpful in the writing of this book; most importantly, the urge to undertake this project was ignited during these meetings.

Thanks are due to Dr. Rambabu Kammili, Director, RCI, Hyderabad, for giving opportunities to give invited talks and attend review meetings on ring laser gyroscopes, thereby providing the chances to interact with the scientists of DRDO Laboratories of Hyderabad and also academicians of the Indian Institute of Science, Bangalore, the discussions with whom helped me in the final refinements of my ideas presented on the topic in the book. Finally, the author is grateful to Prof. Vipin Kumar Tripathi of the Department of Physics, Indian Institute of Technology, Delhi, for various suggestions and encouragement during the course of writing this book, which helped me not only in greatly improving the contents but also in the presentation and readability of the book. The author is thankful to Shri. G. Krishna Rao, Director, Electro Optical Instruments Research Academy (ELOIRA), Hyderabad, for useful discussions, suggestions, and encouragement while finalizing this book.

About This Book

The book is a serious attempt by the author for presenting some useful and important aspects of the optoelectronic gyroscopes: ring laser gyroscopes (RLGs) and fiber-optic gyroscopes (FOGs). The designing aspects for optimizing their performance have been analytically discussed in detail, besides explaining some of the related concepts and the new developments.

Some useful novel designs of RLG given in the literature have also been presented and discussed for the benefit of the optical engineers aiming to design and develop new types of RLG with a view to minimize the size and maximize the longevity of the RLG.

In addition, the related technologies like double ion beam sputtering for fabricating some of the required components like RLG mirrors on the high-quality optical substrates and the optical testing, and thin-film characterization techniques for their evaluation have been discussed in detail at one place. Since the literature about these topics is quite scattered, this book will be able to bridge the gap for the scientists and academicians engaged in working on these topics. Since the quality of the RLG mirrors determines to a large extent the performance of the RLGs, great emphasis has been laid on the design and technology for these mirrors, besides discussing their stringent specifications.

Since the author has hands-on experience on most of the topics presented in the book, he has been able to discuss the minute details and complexities of the subject. Another advantage for the readers is that the book also discusses at length some of the recent experimental results described in the literature and the designing aspects for optimizing the results. In view of this novel combination of concepts, the book is expected to be really useful for the researchers, designers, and engineers working in these high-technology areas of optical gyroscopes and the related technologies.

Contents

About the Author

Dr. Kamal Nain Chopra has done B.Sc. (University of Delhi), M.Sc. (Physics - IIT, Delhi), M.Tech. (Opto-Electronics - IIT, Delhi), and Ph.D. (Applied Physics - IIT, Delhi). He has served DRDO for a period of 33 years and superannuated as Scientist G, from Laser Science and Technology Centre (LASTEC), Delhi, in the year 2005. Subsequently, he has also served as Professor (Physics) in NSIT (DU) and MAIT(GGSIPU), and as Project Scientist in IIT, Delhi, in various Projects, on Topics including Photonics, Thin Films, and Optical Testing.

He has about 390 publications including about 300 in International journals (UK, USA, France, Germany, Italy, Netherland, and China) on various topics including Thin Films Optics, Lasers and Laser Components, Holography, and Modern Optics; 12 Invited talks; 15 Technical reports; and 30 papers in International Conference Proceedings (e.g. Taylor and Francis, UK; and Scientific Net, Switzerland).

Dr Kamal Nain Chopra has co-authored a book titled, "Thin Films and their Applications in Military and Civil Sectors", DESIDOC, DRDO, Ministry of Defence, INDIA, 2010. He has authored a book titled, "Unconventional Lasers: Design and Technical Analysis", DESIDOC, DRDO, Ministry of Defence, INDIA, 2017. He has also authored a Book titled, "Conventional and Unconventional Sources of Renewable Energy: Renewable Energy Sources", Lambert Academic Publishing, LAP, GERMANY, 2017. In addition, he has authored a book titled, "Spintronics Theoretical Analysis and Designing of Devices Based on Giant Magnetoresistance", DESIDOC, DRDO, Ministry of Defence, INDIA, 2019.

He has undertaken visits to foreign universities and industries including (i) School of Thin Film Coatings, Department of Physics, St. Jerome University, Marseille, FRANCE [5 months (1984-85)]; (ii) Department of Physics, Innsbruck University, Innsbruck, AUSTRIA, including 5 days in M/s. Balzers, Liechtenstein, SWITZERLAND [10 days (1995)]; and (iii) M/s. Elettrorava, Torino, ITALY [15 days (2000)].

He has vast experience of serving the Recruitment and Assessment Boards of DRDO (RAC and CEPTAM), as Chairman as well as Expert Board Member. He is a reviewer and editorial board member for some leading international journals.

His fields of specialisation are opto-electronis, unconventional lasers, optical gyroscopes, thin films designing, fabrication, and characterisation by modern techniques, and specialized optical testing techniques.

Chapter 1
Ring Laser Gyroscopes

1.1 Introduction

Various types of sensors are used for determining the position and orientation of an object, the most commonly used being the gyroscope and the accelerometer. The two are similar from purpose point of view, but actually, they measure different things. Interestingly, when combined into a single device, they can serve as a very powerful tool for providing an array of information. Microelectromechanical sensor (MEMS) called MEMS gyroscope measures changes in the forces acting on two identical masses, oscillating and moving in opposite directions. However, their sensitivity is limited, and therefore, optical gyroscopes have been developed, which can perform the same function without any moving parts and much greater degree of accuracy by using a phenomenon called the Sagnac effect.

The Sagnac effect is an optical phenomenon coming from Einstein's theory of general relativity. This is created by splitting a beam of light into two beams, which travel in opposite directions along a circular path and then meet at the same light detector. Light travels at a constant speed, and therefore rotating the device and hence the path that the light travels, results in one of the two beams to arrive at the detector earlier than the other. This phase shift, known as the Sagnac effect, can be used to calculate orientation.

A gyroscope is simply a device which uses Earth's gravity for determining the orientation. In its simple form, the device consists of a freely rotating disk called a rotor, which is suitably mounted on a spinning axis in the center of a larger and heavy, i.e., more stable wheel. The design is such that as the axis turns, the rotor remains stationary, indicating the central gravitational pull.

An accelerometer is a simple device used for measuring non-gravitational acceleration. When the object, it is integrated with, goes from a rest position to any velocity, the accelerometer responds to the vibrations associated with such a movement. A gyroscope is used in an aircraft for helping in indicating the rate of rotation around the aircraft roll axis. When an aircraft rolls, the gyroscope measures nonzero values until the platform returns to original state, when it reads a zero value to indicate the

K. N. Chopra, *Optoelectronic Gyroscopes*, Progress in Optical Science and Photonics 11,
https://doi.org/10.1007/978-981-15-8380-3_1

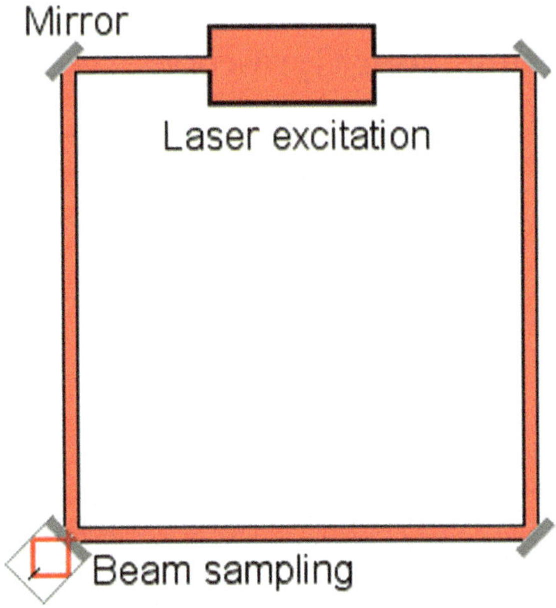

Fig. 1.1 Schematic of a square ring laser setup Figure courtesy ring laser gyroscopes, Wikipedia.org

direction of "down". The schematic of a ring laser gyro (square and triangular) is shown in Figs. 1.1 and 1.2.

The ring laser gyroscopes (RLGs) are very useful as the stable elements in an inertial guidance system, for one degree of freedom for each system. The main advantage of using RLG over conventional mechanical spinning gyroscopes is that it has no moving parts, and hence, there is no friction, which implies that there are no inherent drift terms. Besides, the whole unit is compact, lightweight and nearly indestructible, which qualifies it for use in aircraft and more importantly in missiles. The operation of RLG is based on the fact that a single RLG is capable of measuring any rotation about its sensitive axis, which means that the orientation in inertial space is known at all times. This implies that the elements measuring actual accelerations can be resolved into the suitable directions.

A ring laser gyroscope (RLG) consists of a ring laser having two counter-propagating modes over the same path and is used to detect and measure rotation, the operation of which is based on the principle of the Sagnac effect that shifts the nulls of the internal standing wave pattern in response to the angular rotation. The shifts in the standing wave are shown by the interference between the counter-propagating beams, observed externally. Sagnac interferometer was considered as a possible optical inertial rotation sensor as early as in the year 1913, but could not compete with mechanical gyroscopes, because of being on threshold of experimental resolution, a problem that was solved with the advent of lasers, since it became clear that in an active resonator, it is possible to measure this effect accurately, which ultimately led to the possibility of a ring laser becoming a novel type of gyroscope.

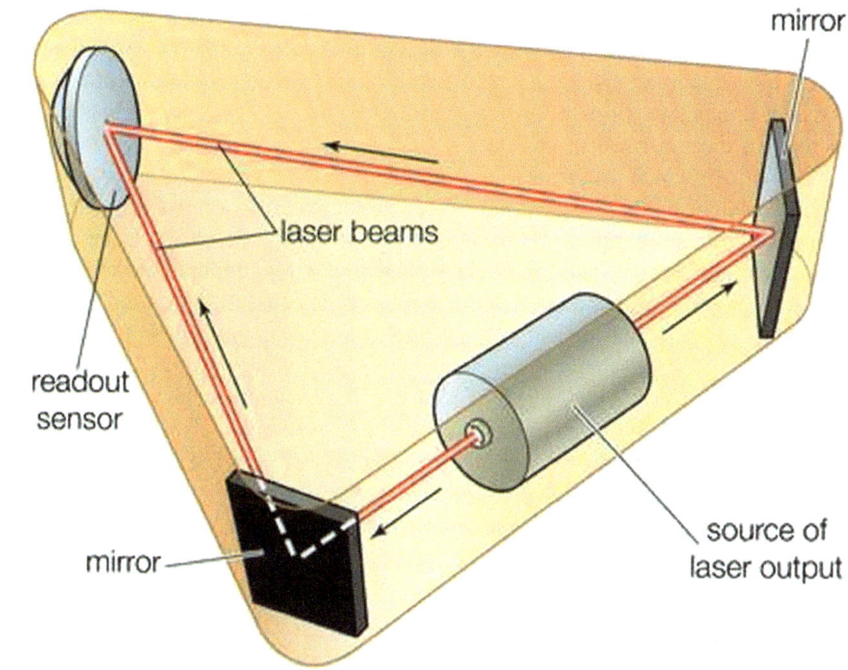

Fig. 1.2 Triangular ring laser gyroscope. Figure courtesy Encyclopædia Britannica, Inc

Many useful investigations have recently been made on the detailed study of the device—RLG, which employs a ring laser to measure rotation, the 0.6328 μm He–Ne laser being universally used, along with the three-mirror equilateral triangle, as the most common resonator geometry. The working principle of the RLG is simple. A ring resonator is used to support two modes propagating in opposite directions, and when the resonator is rotated, the two modes are frequency shifted relatively, the shift being linearly proportional to the rate of rotation. The working of the device is based on heterodyning the two output beams and subsequently measuring their frequency difference.

1.1.1 Frequency Synchronization—Lock-in—of the Two Modes

The problem of lock-in is due to the weak mutual coupling caused by backscattering in the pressure ~ milli Torr, which results in a dead band around zero, and also a strongly nonlinear signal near the threshold. This problem has led to the revolutionary design and development of very low scattering loss RLG mirrors. However, the problem is overcome by biasing technique of back-and-forth rotation, "dithering," which has been adopted and successfully used in many commercial systems. The technique also suffers from random noise, induced by the irregular bias and the loss of information as the system passes through the dead band. Though many methods of biasing have been tried, the most useful has been the four-mirror non-planar resonator with a Faraday mirror which produces four modes widely separate in frequency. However, this technique is really complex and has much nonlinearity because of the presence of so many optical components in the system. Another, useful technique is based on the two modes (i.e., two longitudinal modes or a longitudinal and a transverse one) to oscillate simultaneously: the condition being that the two modes must be approximately of equal intensity, to interact for reducing the lock-in threshold. So much progress has been made in developing the RLGs by various global organizations during the last five decades, as a result of which, it has now been well established that an optically biased no-moving-parts ring laser gyroscope is a very useful device for measuring rotation in the inertial navigation system. A ring laser gyroscope is shown in Fig. 1.3.

The RLG device has three (or four) mirrors arranged in a triangle (or a square), forming a closed path. The operation, based on two counterclockwise beams, is simple and shown in Fig. 1.4.

If the RLG is rotated clockwise, the light having same speed in both the directions takes a bit longer time for the beam in the C direction to reach its starting point than the one in the CC direction, because the starting point itself has moved in this interval, and this results in the path difference between the two beams, which is measured, and converted into a digital output. The counter-propagating laser beams

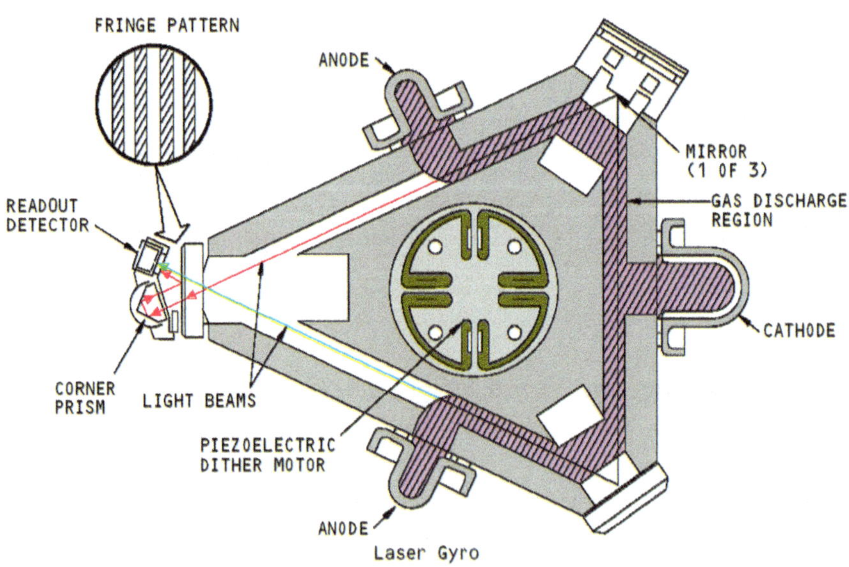

Fig. 1.3 Schematic of RLG setup. Figure courtesy www.k-makns.gr

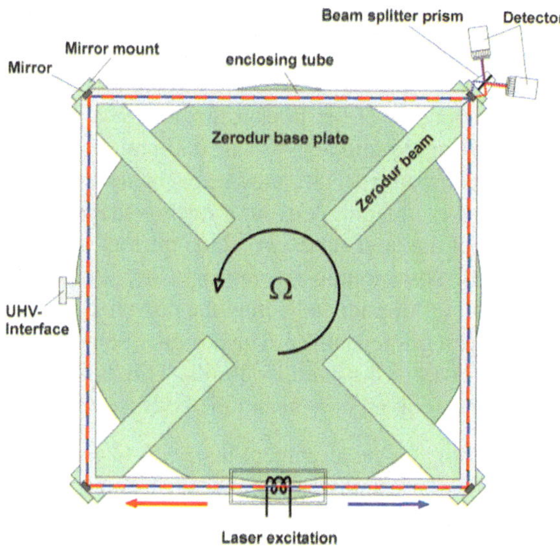

Fig. 1.4 Principle of operation of RLG. Figure courtesy www.wittzell.ifag.de

have different frequencies with the difference being dependent on rotation rate, and the measurement of this difference gives this rotation rate about the RLG's sensitive axis and thus makes it possible to determine the orientation of a system in inertial space at all times. The rate of rotation is measured by a computer, the output system for measuring the angular rotation being dependent on the generation of the interference patterns in the light output.

It may be noted that the output is taken out of the laser cavity through the output mirror of transmission of about 0.05%. At present, ring laser gyroscopes are being routinely used as the stable elements for one degree of freedom each, in an inertial navigational system, because of its advantage of not having any moving parts, other than the dither motor assembly, thereby being free from friction and hence eliminating a considerable amount of drift. In addition, it has many other advantages like the compactness of the full unit, lightweight, highly durable, and having no resistance to any changes to its orientation, and consequently, it is suitable for use in aircraft, missiles, and satellites. Interestingly, the ring laser gyroscopes (RLGs) are also found useful ss embedded GPS capability for further enhancing the accuracy of RLG inertial navigation systems (INSs) on military and commercial aircraft, and spacecraft.

1.1.2 Alignment of Optical Elements in RLG

Alignment of the optical elements is critical for the successful and optimized performance of RLG and must be done very carefully, since the He–Ne laser works only when the alignment is very precise. For giving some idea, this process is briefly discussed here. This requires great skill on the part of the experimentalist. Though some instructions for getting a good alignment are known, still a lot of patience for following the trial-and-error technique is required. It is absolutely important that the experimentalist starts the next step only after completing the ongoing step. The components of the laser resonator (Sec. 1.1.1, Fig. 1.3) are He–Ne laser with adjustable holder, diode laser with collimation optics, two plane mirrors with mounts, one spherical mirror with mount, beam splitter cube with mount, two photodiodes with mounts, imaging lens with raising adapter. The resonator setup to be aligned consists of the laser tube, two plane mirrors (say M1 and M2), and a curved mirror (say M3). For aligning the mirrors, first, the alignment laser is coupled into the resonator, followed by adjusting the mirrors until the alignment laser interferes with itself after a full revolution. The designer of the setup has to ensure that the resonator length is within the stability limits. This is followed by setting up the two plane mirrors, the curved mirror, and the laser tube, according to the steps given below:

(i) Turning on the alignment laser,
(ii) Adjusting the laser tube until the beam from the adjustment laser goes exactly through the laser tube's capillary. Then, a series of steps like exchanging mirror M2 with the imaging lens and putting a screen behind, putting M2 back and adjusting it, so that the beam falls exactly on the center of the curved mirror M3, adjusting M3 so that the beam falls on M1 exactly where the beam passes through the first time, adjusting M1 for the beam to be directed through the tube again.

It is important to mention here that the last step is very tedious and difficult, because it is very hard to see the beam after the second pass through the tube. Then by manipulating M1and M3, the secondary beam has to be adjusted until some interference can be observed on the screen. This is followed by careful adjustment of the same two mirrors, for improving this interference pattern so that it is roughly concentric. To make the things easier, mirrors M1 and M2 must be adjusted either horizontally or vertically at a time, and they should be adjusted in the direction, which provides the larger distance between maxima (or minima) of the interference pattern. The final steps are positioning mirror M4 (mirror for final setting and verification) so that the two outgoing beams cross at a right angle, positioning the beam splitter so that the two beams cross at the very center of the cube and also enter the cube at a normal angle with its faces, and adjusting the positions of mirror M4 and beam splitter so that the two beams overlap. The perfection of the alignment is verified by positioning the imaging lens behind the beam splitter and shifting both the beam splitter and the mirror simultaneously, in the same plane, so that the beams are constantly overlapping.

Principle of Operation: As explained earlier, the rate of rotation produces a small difference between the times taken by the light to traverse the ring in the two opposite directions due to the Sagnac effect, which results in introducing a small separation between the frequencies of the counter-propagating beams in the form of a motion of the standing wave pattern inside the ring, and subsequently, a beat pattern on combining these two beams to interfere outside the ring. So, it is clear that the net shift of this interference pattern is a function of the rotation of the unit in the plane of the ring.

RLGs are designed in different shapes—triangles, squares, or rectangles—and filled with inert gases, through which the beams are reflected by mirrors, which, though working like laser mirrors, are very special and made with difficulty, so as to be having nearly negligible scattering loss to avoid the lock-in problem, besides having negligible absorption loss for efficient performance of the RLGs. These mirrors are so difficult to design and develop, which makes them very expensive and critical component of the device. It is not surprising that only about three companies including M/s OCLI and M/s Oxford are able to make them, though they do not sell these to the organizations engaged in making RLGs, and so, they have to make these in order to make good RLGs of high quality. Special polishing techniques for making the optical substrates like float polishing and sophisticated coating techniques like dual ion beam sputtering are required along with very highly skilled and experienced technical personnel. In addition, sophisticated optical testing (like differential contrast microscope) and thin films characterization instruments (like atomic force microscope and cavity loss meter) are required. The next step is to measure the interference patterns created in the corresponding rings of the gyroscope, as the vehicle undergoes a turning or pitching motion, which is done by using the photoelectric cells. Finally, the patterns of all the three rings are numerically integrated to determine the turning rate of the craft in three dimensions.

1.2 Optimization and Mathematical Modeling of the RLGs

It is important to note that the Sagnac effect is independent of the choice of reference frame, which can be understood from the simple calculation, invoking the line element of the resultant metric is given by:

$$ds^2 = (c^2 - r^2\Omega^2)dt^2 - dr^2 - r^2 d\theta^2 - dz^2 - 2r^2\Omega dt d\theta, \tag{1.1}$$

where t is proper time for the central observer, r is the distance from the center, θ is the angular distance along the ring from the direction faced by the central observer, z is the direction perpendicular to the plane of the ring, and Ω is the rate of rotation of the ring and the observer. In this case, the speed of light tangent to the ring is $(c \pm r\Omega)$ depending on whether the light is moving against or along the rotation of the ring, $\Omega = 0$ being the only inertial frame of reference, and $\Omega \neq 0$ being the non-inertial frame of reference, which explains as to why the speed of light at positions distant from the observer (at $r = 0$) can have any value other than c. Interestingly, the Sagnac effect is being used in the current technology, in case of the inertial guidance systems. RLGs are extremely sensitive to rotations, which have to be measured accurately for the successful performance of the inertial guidance system.

1.2.1 Enhancement of the Efficiency of the RLGs

For the efficient performance of the RLGs, ideally, the optics must be very well isolated from the seismic motion of the ground, e.g., at ~10 Hz, and an isolation of 9 orders of magnitude is needed for achieving high degree of design sensitivity [1]. This is done by changing the interferometer length away from the correct operating point, at the lower frequencies, which results in the nonlinear coupling of mirror motion to output signal, and at the extremes of this, the detector is free from lock-in. In fact, the low frequency isolation is achieved actively, by using the seismometers for feeding forward to the hydraulic HEPI actuators. It is observed that the sensitivity of the seismometers to the horizontal motion at low frequencies is sufficient for achieving the required isolation. The problem is that the coupling of rotations, or ground tilt, into horizontal seismometer signals at low frequency is very difficult. This can be understood in a simple way explained below.

For a horizontal seismometer, the ratio of (i) sensitivity to rotation to (ii) sensitivity to horizontal motion at a frequency ω is given by:

$$\left[\frac{\text{rotation sensitivity(RS)}}{\text{horizontal sensitivity(HS)}}\right] = -\left[\frac{g}{\omega^2}\right]. \tag{1.2}$$

It may be understood that the response of the seismometers is expected to be dominated by tilt, below some frequency. Hence, if this signal is fed forward into the system, it can produce the horizontal translations in response to these wrong signals, which can, however, be avoided by using a rotation sensor in parallel with the seismometers. The rotational sensitivity required is calculated by assuming that the noise from the rotational sensor is ~1/10th of the total noise in the horizontal direction, and hence, by using the above equation, we can express this as given below:

$$\Omega_{\text{sensitivity}} = \frac{1}{10}\frac{\omega^2}{g}x_d, \tag{1.3}$$

where x_d is the horizontal sensitivity requirement, which, for 0.2 Hz, comes out to be

$$x_{\text{d}} = 3 \times 10^{-9}\text{rad}/\sqrt{\text{Hz}}. \tag{1.4}$$

These values help the designer in optimizing the rotational sensitivity of the RLG, which is obviously improved by lowering the value of the ratio of the rotational sensor noise to the total noise. Though the efforts are continuously going on improving the value of x_{d}, in general, this is in the range as given below:

$$x_{\text{d}} = 3 \times 10^{-9}\text{rad}/\sqrt{\text{Hz}} \quad \text{to} \quad x_{\text{d}} = 10^{-10}\text{rad}/\sqrt{\text{Hz}}$$

1.2.2 Externally Excited Laser Gyroscope Technology

As already mentioned, the RLGs operate on the Sagnac principle, in which the path length for light traveling around a ring is changed on its rotation, and hence, the counter-propagating beams interfere at the output, providing a beat frequency, which is proportional to the rotation rate. This fact is used in many applications, the most important being the RLGs of various sizes (e.g., 16 and 800 m^2). In case of a passive ring gyroscope using fixed mirrors, the two counter-propagating beams are locked to a triangular cavity, shown in Fig. 1.5.

The counterclockwise beam (CCW) is modulated by using an electro-optic modulator (EOM) to produced sidebands, which allow Pound–Drever–Hall (PDH) locking [2] to the cavity by changing the laser frequency, and in the same way, the CW beam is modulated and locked using an acousto-optic modulator (AOM) to shift the frequency. An acousto-optic modulator, also called a Bragg cell, uses the acousto-optic effect to diffract and shift the frequency of light using sound waves (usually at radio-frequency) and is used in lasers for Q-switching, telecommunications for signal modulation, and spectroscopy for frequency control. It has to be understood that at the beam sampling location, a fraction of each of the counter-propagating beams is made to exit the laser cavity. RLGs are used as the stable elements in an inertial navigation system, one RLG serving for one degree of freedom each. The advantage of using an RLG over the conventional spinning gyroscope is that apart from the dither motor assembly, there are no moving parts, and hence no friction, which implies that the system is free from any inherent drift terms. Also, the entire

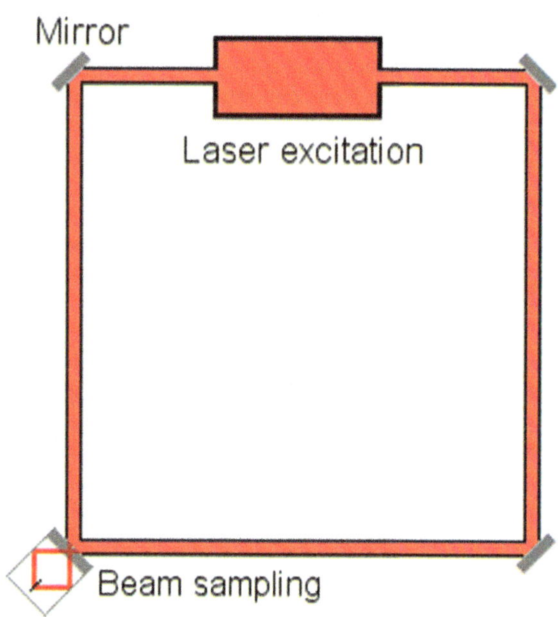

Fig. 1.5 Laser excitation and beam sampling in RLG. Figure courtesy wikimedia.org

unit being compact, lightweight, and virtually indestructible with no resistance to its orientation is really suitable for use in the aircraft. The beat frequency between the modulation frequencies is a function of the rotation rate, Ω, and is given by:

$$\Delta f = \frac{4A}{\lambda P}\Omega. \tag{1.5}$$

where A is the area of the ring, P is the perimeter of the ring, and Ω is the rotation rate.

The lock-in problem [3] is avoided by using different modulation frequencies for each of the two counter-propagating beams. The sensitivity $\delta\Omega$ of a shot noise limited passive laser gyroscope [4] is given by:

$$\delta\Omega = \left(\frac{\lambda P}{4A}\right)\frac{\sqrt{2}\Gamma}{\sqrt{(n_{ph}\eta\tau)}}, \tag{1.6}$$

where Γ is the bandwidth, n_{ph} is the number of photons arriving at the detector, η is detector efficiency, and τ is the integration time. Just to have an idea about the sensitivity, for a triangular cavity with 1 m sides and finesse of 1000, the sensitivity achieved is

$$\sim 1 \times 10^{-10}\text{rad}/\sqrt{\text{Hz}}.$$

Thus, $\delta\Omega$ depends on various parameters—P, A, Γ, n_{ph}, η, and τ. These values have to be optimized, so that a workable RLG of proper dimensions (and therefore of proper weight) and of the required value of $\delta\Omega$ can be made available, by adjusting the values of Γ, n_{ph}, η, and τ. This requires a lot of skill on the part of the designer and fabricator of the RLG.

1.2.3 The Optical Path Difference and the Corresponding Phase Difference

Considering the circular path (Radius R), the path lengths for the two counterclockwise beams are given by the following equations:

$$ct_+ = 2\pi R + R\Omega t_+ \Rightarrow t_+ = \frac{2\pi R}{c - \Omega R}. \tag{1.7}$$

and

$$ct_- = 2\pi R - R\Omega t_- \Rightarrow t_- = \frac{2\pi R}{c + \Omega R}. \tag{1.8}$$

The optical path difference and the corresponding phase difference are given by the following equations:

$$\Delta L = c(t_+ - t_-) \approx \frac{4\pi R^2 \Omega}{c}. \tag{1.9}$$

and

$$\Delta\phi = 2\pi \frac{\Delta L}{\lambda} = \frac{8\pi}{\lambda c} A\Omega. \tag{1.10}$$

For an N-sided regular polygon, these equations are modified as given below:

$$\Delta L = 2\frac{2\sum A\Omega}{c} = \frac{4A_{\text{enc}}\Omega}{c}, \tag{1.11}$$

and

$$\Delta\phi = \frac{8\pi}{\lambda c} A \cdot \Omega, \tag{1.12}$$

$$A = \frac{1}{2}\oint r \times \mathrm{d}r. \tag{1.13}$$

where c is the velocity of light., and A is the area of the polygon, given by:

$$A = \frac{1}{2}\oint r \times \mathrm{d}r$$

In the RLG, an active laser medium is introduced into the cavity, which effectively converts phase changes to frequency changes, given by:

$$L = m\lambda = m\frac{c}{\nu}, \tag{1.14}$$

and

$$\delta\nu = \frac{\nu\delta L}{L} = \frac{c}{\lambda L}\delta L = \frac{c}{\lambda L}\left(\frac{\lambda}{2\pi}\Delta\phi\right) = \frac{4\vec{A}\cdot\vec{\Omega}}{\lambda L}. \tag{1.15}$$

Finally, the intensity is modulated at the beat frequency, as given below:

$$I = I_o(1 + \cos(2\pi\Delta\nu t)). \tag{1.16}$$

Just to have some idea about the practical value, the computations show that, for

$d = 10$ cm, $A = 45$ cm^2, $l = 633$ nm, and $\vec{\Omega} = 15$ deg/h $= 7.3*10–5$ rad/s, $\Delta\phi$ comes out to be $4.2*10^{-8}$ rad. Thus, the designer has to choose the value of $\Delta\phi$ for the values of d and A (corresponding to a workable RLG), required for measuring the particular rotation rate.

Representing the time difference for the two counter-rotating beams in Eq. (1.9), by Δt, we can write the following expression:

$$(t_+ - t_-) = \Delta t = \left\{\frac{4\pi R^2 \omega}{c^2}\right\} = \left\{\frac{4A\omega}{c^2}\right\}, \tag{1.17}$$

and so, the phase shift can be expressed as:

$$\Delta\phi = \left\{\frac{2\pi c \Delta t}{\lambda}\right\}. \tag{1.18}$$

Thus, it is clear that the fringes are shifted in proportion to A and ω. The designer has to take these factors into consideration while choosing the value of A and the related parameters.

1.2.4 Noise Due to Optical Kerr Effect

There can be several sources of noise. The optical Kerr effect is very important source of noise, since the electric fields of the counter-propagating beams cause changes in the index of refraction, which is non-reciprocal if $|E_1|$ and $|E_2|$ are not equal, e.g., $I_1/I_2 = \pm\ 10^{-4}$ leads to 10^{-3} deg/h shifts.

$$\upsilon_{\mathrm{CW}} = \frac{c}{[\varepsilon_o + \varepsilon_2(|E_1|^2 + 2|E_2|^2]^{1/2}}, \tag{1.19}$$

and

$$\upsilon_{\mathrm{CCW}} = \frac{c}{[\varepsilon_o + \varepsilon_2(|E^2|^2 + 2|E_1|^2]^{1/2}}. \tag{1.20}$$

This problem is usually eliminated by doing the square wave modulation, as shown in Fig. 1.6.

It has been observed that the 10^{-3} deg/h shifts can be easily eliminated by the square wave modulation.

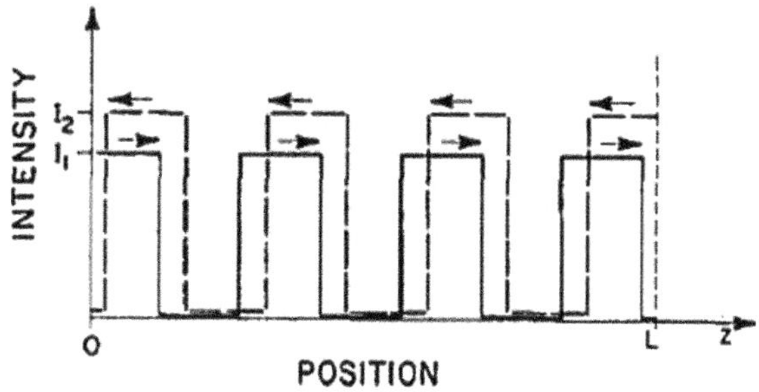

Fig. 1.6 Square wave modulation to eliminate the optical Kerr effect

1.2.5 Effect of Null Shift, and Mode Locking on the Performance of the RLGs

There are many types of sources of error including null shift and mode locking. The null shift arises due to the fact that the beat frequency is nonzero even when $\Omega = 0$. This is explained on the basis of the Langmuir flow, according to which, in the active laser media, neutral atoms along the center of the discharge move toward the cathode, while the atoms near the walls move toward the anode. Because of the fact that the lasing light is concentrated in the center, the two counter-propagating beams see opposite motion for the lasing medium, and hence, different indexes of refraction are termed as Fressnel drag. This is usually overcome by using two discharge tubes. The mode locking is due to the fact that the backscattering in the optical path (mirrors) weakly couples the counter-propagating beams, and thus in case, the beat frequency gets smaller than a threshold value, and the modes oscillate at the same frequency, eliminating the beat note. The equations for $\dot{\psi}$, *S*, and ψ are given below:

$$\dot{\psi} = S\Omega + b \sin \psi, \tag{1.21}$$

where

$$S = \frac{4A}{L\lambda}. \tag{1.22}$$

and

$$b = \frac{\pi \nu \chi_0 r}{\varepsilon_0}. \tag{1.23}$$

For $S\Omega < b$, there is a stable solution where

$$\psi = \pi + \mathrm{arc}\, sin\left(\frac{S\Omega}{b}\right). \tag{1.24}$$

The computations show that Ω_{TH} » $b/S = 400$ deg/h for $b = 10^3$ rad/s.

The mode locking is avoided by applying: (i) constant bias and (ii) alternating bias. In case of the constant bias, a large constant bias (a) is added such that within the operating range, $(S\Omega + a) < b$. The problem is that of keeping a large bias very stable and also knowing its precise value, so as to be subsequently able to subtract it. Use is also made of the Faraday effect, in which a magnetic field is applied parallel to the direction of propagation that leads to polarization rotation given by:

$$\beta = VBd,$$

where β is the angle of rotation (in radians), B is the magnetic flux density in the direction of propagation (in teslas), d is the length of the path (in meters) where the light and magnetic field interact, and V is the Verdet constant (empirical proportionality constant) for the material. However, the rotation is the same for light in either direction leading to a frequency shift:

$$\Delta\nu = \frac{c}{\lambda L}\left(\frac{\lambda}{2\pi}\Delta\phi\right) = \frac{c}{\pi L}VBd. \tag{1.25}$$

It should be noted that the Verdet constant for the material has units of radians per tesla per meter and varies with wavelength and temperature.

In case of the second technique, the alternate bias is applied in the positive and negative directions, and because of the fact that over each cycle the net bias averages out, the need to know the precise value of the amplitude is eliminated. The phase is given by:

$$\dot{\psi} = S\Omega + b\,\sin\psi + \alpha\,\cos(\omega t)a,\, w \gg b. \tag{1.26}$$

By solving this differential equation, we can see that there are "dead bands" centered at the multiples of the dither frequency with width:

$$J_{-r}(\tfrac{\alpha}{\omega}) \cong (\tfrac{2\omega}{\pi\alpha})^{1/2}\cos(\tfrac{\alpha}{\omega} + \tfrac{r\pi}{2} - \tfrac{\pi}{4}). \tag{1.27}$$

The dead spaces can be made very small by choosing very large value of $\frac{\alpha}{\omega}$, which can be done by increasing α and decreasing ω, which are very important indeed for the optimizing the performance of the RLG.

Some results for the ring laser gyroscope without the lock-in phenomenon have been obtained by Sunada et al. [5] and have been reproduced in Fig. 1.7.

As is expected, the curve shows a nearly linear response of the frequency difference to of the refractive index. Besides these sources of error, the amplification of the light wave takes place during its passage through the gain medium, the shape of the gain curve being determined by a combination of (i) homogenous and (ii) inhomogeneous broadening. Homogenous broadening affects all the atomic dipoles uniformly and is of two forms: lifetime broadening, which occurs because of spontaneous emission of the lasing energy level, and the collision broadening, which is

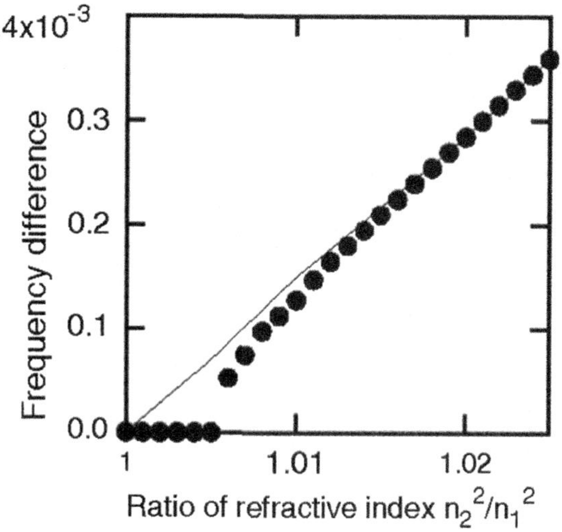

Fig. 1.7 Frequency difference as a function of the ratio of refractive index $\{\frac{n_2^2}{n_1^2}\}$. Figure courtesy Sunada et al. [5]

due to collisions between atoms, ions, free electrons, and the walls of the gain tube. In case of the lifetime broadening, the Heisenberg uncertainty principle provides a minimum uncertainty in the energy of the transition, which, for the neon transition (5 s to 3p) at 632.8 nm, has the decay time ~55 ns for the upper level and ~7–10 ns for the lower level [6]. The natural broadening leads to a Lorenzian distribution, and the full width half maximum of the Lorentzian line is given by the equation below:

$$\gamma = \left\{\frac{1}{2\pi\,\tau_{\text{upper}}}\right\} + \left\{\frac{1}{2\pi\,\tau_{\text{lower}}}\right\}, \tag{1.28}$$

where τ is the spontaneous decay rate of the energy level, and for the transition given above, the computed value comes out to be 18 MHz. In case of the collision broadening, the atoms, ions, and free electrons move in a random manner because of the thermal and Brownian motions. The collision occurrence causes a disturbance in the phase of the electric field of the species undergoing the collision. The homogenous broadening increases linearly with total pressure of He and Ne, and for a He–Ne mixture, this broadening is ~77 MHz/Torr [7]. Inhomogeneous broadening is considered to be the most important broadening mechanism for gas lasers, which occurs because of atoms having different resonance frequencies, randomly distributed from a central frequency. Doppler broadening is found to be the dominant process in broadening for He–Ne lasers, which is caused by Doppler shifts of the individual atoms because of the thermal motion, and the Doppler broadening width $\Delta\nu$ is given as:

$$\Delta\nu = 2\nu_0\left\{\frac{2k_B T\,\ln(2)}{Mc^2}\right\}^{\frac{1}{2}}, \tag{1.29}$$

where ν_0 is the lasing frequency of 4.74×10^{14} Hz, k_B is Boltzmann's constant, T is the temperature, and M is the mass of Ne. The computations show that its value is 1.5 GHz which is in complete agreement with the quoted value of 1.5 GHz [8].

1.2.6 RLG Using Multilayer Optical Coatings with Huge Group Delay

It is well known that an actual account of the angle random walk (ARW) coefficients of gyros in the constant rate biased rate ring laser gyro inertial navigation system (RLG INS). Yu et al. [9] have discussed that in this system, an advantage is that the validity of the FOS-based method can be checked by estimating the ARW coefficients of the mechanically dithered RLG under stationary and turntable rotation conditions. It has been emphasized that by utilizing the FOS-based method, the average ARW coefficient of the constant rate biased RLG in the postulate system can be estimated. An added advantage of the method, as shown by the experimental results, is that the FOS-based method can achieve high denoising ability. It is now well established that the RLGs are the industry standard for precision rotation measurement, since they are the critical sensing elements in practically all types of the navigation systems for manned and unmanned aircraft, platform stabilization systems, pointing and targeting, marine navigation and attitude and heading reference systems (AHRS). This is because of the fact that the RLGs have the advantage of precision rotation measurement at a low cost; with no moving parts which lead to improve significantly the long term reliability; and a very compact form factor. Qu et al. [10] have proposed and analyzed a superluminal ring laser gyroscope using multilayer optical coatings with huge group delay (GD). Their results of the beat frequency f_b of both fast-light enhanced (central curve) and standard RLGs (straight line) with respect to the angular rotation rate Ω_r are shown in Fig. 1.8.

Fig. 1.8 Beat frequency f_b of both fast-light enhanced (central curve) and standard RLGs (straight line) with respect to the angular rotation rate Ω_r. Full size image (197 KB). Figure courtesy Qu et al. [10]

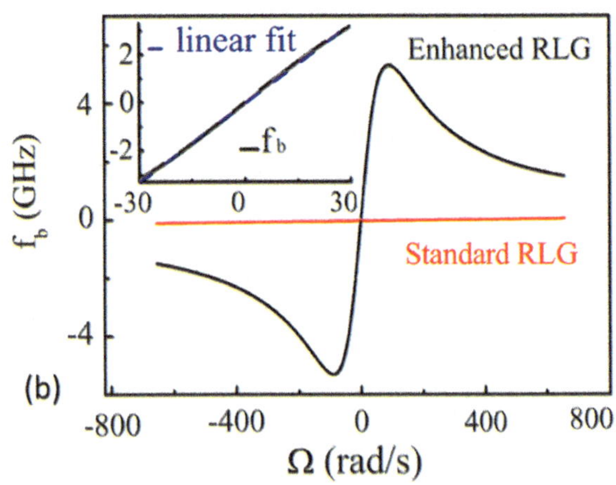

Apart from the technique of solving the lock-in problem, a lot of effort has gone into making the RLG mirrors of very low scattering loss ~5–10 ppm level [11], by using dual ion beam sputtering unit [12], which helps even in increasing the damage threshold and the life of the mirrors, and so of the RLG. Also, total internal prisms of very high optical quality are employed in place of mirrors to avoid this problem. In fact, all the steps and techniques are applied together to achieve the desired results. Efforts are in progress for improving the quality of the surface of the substrates for the mirrors and also of coating techniques for the fabrication of the mirrors. Cho et al. [13] have presented the Design and Development of an Ultralow Optical Loss Mirror Coating for Zerodur Substrate, and Stover [14] has given the relation between the surface as roughness in rms and total integrated scattering TIS, which is expressed as:

$$\mathrm{TIS} = \left\{\frac{4\pi\delta}{\lambda}\right\}^2, \tag{1.30}$$

where δ is rms surface roughness of the measured surface and λ is the wavelength of the incoming light. Thus, reduction in δ leads to a manifold reduction in the scattering, which has actually led to the better polishing techniques like float polishing for the fabrication of the substrates and state-of-the-art deposition technique of double ion beam sputtering (DIBS) units for making the RLG mirrors.

1.3 Design of Superluminal RLG with ML Optical Coatings

Recently, the topic of RLGs with enhanced sensitivity has drawn the attention of researchers [15–18]. The superluminal RLG is redesigning of the commonly used RLG without much reengineering and has the advantage of miniaturization of size, based on theoretical computation. The principle of application of GD coatings as a fast-light medium is commonly used in the designing of fast-light sensors. Controlling the speed at which light propagates has been the subject of attention for the researchers for more than the last two decades. Controlling the speed of light propagation is fundamentally very important, and the tunability of light speed has led to many diverse applications including optical data buffering and enhanced precision in interferometry. In addition, it has been observed that a fast-light medium can be used to realize an absolute rotation sensor whose sensitivity is enhanced by a factor of 10. Interestingly, the fast-light enhanced gyroscope is able to detect the gravitational frame-dragging effect terrestrially by measuring the Lense–Thirring rotation, and the enhancement is being induced by a frequency-dependent phase shift within a ring laser gyroscope (RLG). This enhancement effect has been shown both theoretically and experimentally by means of increased "mode pushing" or "mode pulling" effects. Many systems have been studied for achieving optimal performances for superluminal gyroscope applications, e.g., alkali metal vapor cells, coupled optical resonators, photorefractive crystals, optical fibers (dual-pumped Brillouin gain in a fiber or fiber-coupled whispering gallery resonators), and spectral hole burning, and rare atomic gases.

Sensitivity enhancement factor S_{enh} superluminal RLG.

Superluminal RLG has been found to give an enhanced sensitivity factor and is important for understanding its advantage over normally used RLG. As we know, for an active RLG, the clockwise (CW) and counterclockwise (CCW) ring laser modes have different frequencies given by:

$$(\Delta f = f_{-} - f_{+}),$$

due to the difference in the effective round-trip optical path lengths resulting from the rotation of the cavity. Following the approach of Qu et al. [15], and denoting $f \pm$, $\lambda \pm$, and $\langle L \pm \rangle$ as the frequencies, wavelengths and the effective optical cavity lengths covered by the CW and CCW propagating beams, respectively, $\langle L \pm \rangle$can be written in the following form:

$$\langle L+\rangle = \frac{\langle L\rangle}{1-\frac{r_0\Omega}{c}} \quad \langle L-\rangle = \frac{\langle L\rangle}{1+\frac{r_0\Omega}{c}}, \tag{1.31}$$

where $\langle L\rangle$ is the round-trip optical path length of the RLG at rest, r_0 is the radius of the beam path for ring cavities, Ω is the angular velocity of rotation about the normal axis through the center of the interferometer, and c is the speed of light. If we take into consideraton the large GD effects of fourth M, the frequency difference Δf induces additional phase difference, and hence, the resonance conditions for the CW and CCW propagating beams may be expressed as:

$$\begin{cases} 2\pi \cdot \frac{\langle L+\rangle}{\lambda_+} - \alpha \frac{\mathrm{d}\phi}{\mathrm{d}f}\Delta f = 2q_2\pi \\ 2\pi \cdot \frac{\langle L-\rangle}{\lambda_-} + (1-\alpha)\frac{\mathrm{d}\phi}{\mathrm{d}f}\Delta f = 2q_2\pi \end{cases}, \tag{1.32}$$

where α is a factor with a value between 0 and 1, q_1 and q_2 are integers, and $\lambda_+ = c/f_+$, $\lambda_- = c/f_-$. Combining Eqs. (1.1) and (1.2), we can obtain the following equation:

$$\frac{\langle L\rangle}{1-\frac{r_0\Omega}{c}} \cdot \frac{2\pi f_+}{c} - \frac{\langle L\rangle}{1+\frac{r_0\Omega}{c}} \cdot \frac{2\pi f_-}{c} - \frac{\mathrm{d}\phi}{\mathrm{d}f}\Delta f = 2m\pi, \tag{1.33}$$

where $m = (q_1 - q_2)$. If we assume that $m = 0$ and $r_0\Omega \ll c$, and use Taylor series and the first-order terms, we can rewrite Eq. (1.33) as:

$$\langle L\rangle \cdot \left(1+\frac{r_0\Omega}{c}\right) \cdot \frac{2\pi f_+}{c} - \langle L\rangle \cdot \left(1-\frac{r_0\Omega}{c}\right) \cdot \frac{2\pi f_-}{c} - \frac{\mathrm{d}\phi}{\mathrm{d}f}\Delta f = 0. \tag{1.34}$$

By this approximation, we have $(f_+ + f_-) = 2f_O$, where f_O is the frequency of the CW and CCW beams in RLG without rotation. Thus, we can get the following result:

$$\Delta f = \frac{4A}{\langle L\rangle\lambda}\Omega\cdot\frac{1}{\left(1+\left(\frac{\mathrm{d}\phi}{\mathrm{d}\omega}\right)/\left(\frac{\langle L\rangle}{c}\right)\right)}, \tag{1.35}$$

where

$$A = \frac{1}{2}Lr_0$$

is the area enclosed by the beam path, λ is the wavelength of the CW and CCW beams in absence of rotation. For

$$GD = -\frac{\mathrm{d}\phi}{\mathrm{d}\omega} = 0, \quad \Delta f = \frac{4A}{\langle L\rangle\lambda}\Omega,$$

Equation (1.35) becomes the formula for the common RLG.

For

$$GD = -\frac{\mathrm{d}\phi}{\mathrm{d}\omega} \neq 0,$$

a sensitivity enhancement factor S_{enh} for the superluminal RLG may be computed as:

$$S_{\mathrm{cnh}} = \frac{1}{\left(1+\left(\frac{\mathrm{d}\phi}{\mathrm{d}\omega}\right)/\left(\frac{\langle L\rangle}{c}\right)\right)}. \tag{1.36}$$

Thus, it is clear that when

$$-1 < \left(1+\left(\frac{\mathrm{d}\phi}{\mathrm{d}\omega}\right)/\frac{\langle L\rangle}{c}\right) < 1,$$

$|S_{\mathrm{enh}}| > 1$, and the scale factor of RLG is enlarged. For

$$1+\left(\frac{\mathrm{d}\phi}{\mathrm{d}\omega}\right)/\left(\frac{\langle L\rangle}{c}\right) = 0, \quad \mathrm{GD} = \frac{\langle L\rangle}{c},$$

the enhancement factor reaches its maximum value. Therefore, the designer has to design multilayer optical coatings with GD around $\frac{\langle L\rangle}{c}$ and at the same time, maintaining the high-reflectivity property of the coatings for RLG. It is observed that for multilayer optical coatings, many designs meet these criteria. However, for practical applications, the commonly chosen quarter-wave multilayer structure is as given below:

$$\text{Substrate}/(\text{HL})^{25}\text{H2L}(\text{HL})^{14}/\text{Air}, \qquad (1.37)$$

where H and L denote the quarter-wave optical thickness layers with high and low refractive index at 45° angle of incidence at the central wavelength of 632.8 nm. The following figure shows the theoretical refractive index profile of a high-reflectivity Ta_2O_5–SiO_2 multilayer coatings (G/(HL)25H2L(HL)14/A). It is to be noted here that in the case of dual ion beam sputtering (DIBS) for the RLG mirrors, Ta_2O_5 is mostly selected as the high refractive index material ($n = 2.125$) and SiO_2 as low refractive index material ($n = 1.46$). Figure 1.9a shows the multilayer structure of G/(HL)25H2L(HL)14/A, composed of a 23-layer high-reflectivity mirror ((HL)11H) and a 57-layer narrow bandpass filter ((LH)142L(HL)14), and Figs. 1.9b, c show the computed reflectivity and GD curves as functions of wavelength for the RLG design illustrated in Fig. 1.9a. It can be observed that there is a broadband high-reflectivity

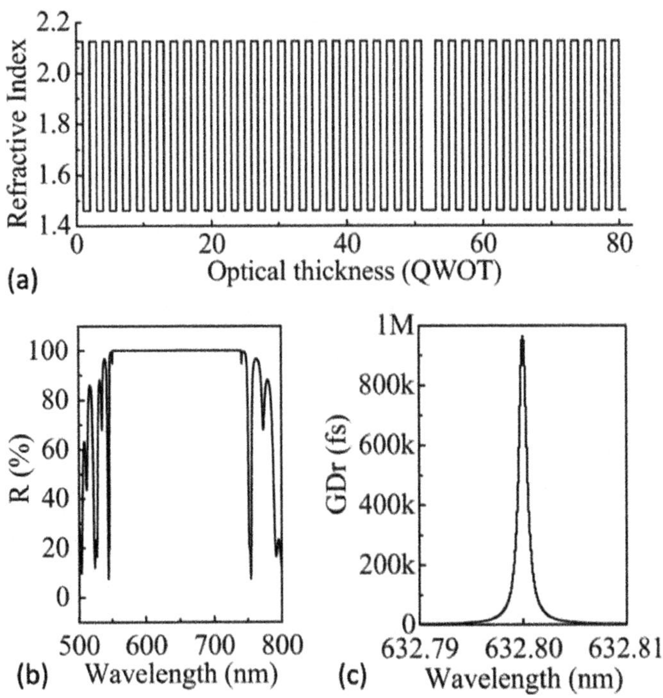

Fig. 1.9 **a** Theoretical refractive index profile of a high-reflectivity Ta_2O_5–SiO_2 multilayer coatings (G/(HL)25H2L(HL)14/A) with huge group delay for M4. **b, c** Computed reflectivity and group delay as a function of wavelength for the multilayer design discussed above. Figure courtesy Qu et al. [10]

multilayer-coating system with huge GD (964,228.61851 fs at maximum) at the central wavelength of 632.8 nm. At the maximum GD, for the cavity optical length of 289.55 mm, the enhancement factor S_{enh} can reach as high as 1029 according to Eq. (1.8). Both the high-reflectivity mirrors in RLG and the bandpass filters in dense wavelength division multiplexing (DWDM) are widely used, and the technology is mature; therefore,this kind of superluminal RLG should be achievable.

1.4 Design of Optimal Degaussing Electronics for Ring Laser Gyroscope

Chelli et al. [19] have discussed that the ring laser gyros are effective tools for large-scale geodetic surveying at a high level of accuracy, as they allow rotation of a sensor block with the system and analytical transformation of the output to the coordinate frame of interest such as frequency difference between two oppositely directed laser beams. Chelli et al. [19] have presented the technique of elimination of the remnant magnetic field in a rectangular coil carrying current efficiently at some point of time by using optimal analysis. They have been able to eliminate the remnant magnetic field, by the method called degaussing, which can be achieved by PWM by changing the duty cycle in different ways. They have further discussed the design of optimal degaussing electronics by optimization principle comprising of selecting one of the three degaussing models.

Process:

Degaussing is the technique of decreasing or eliminating an unwanted magnetic field (named after Carl Friedrich Gauss, a renowned researcher in the field of Magnetism). Because of the magnetic hysteresis, it is not possible to reduce the magnetic field completely to zero, so degaussing induces a very small "known" field referred to as bias. As discussed by Chelli et al. [19] for the degaussing purpose, a code is generated and fused that generates the code into an IC. After dumping the code into the integrated circuit, it is fixed to the degaussing PCB and a kind of wave form is generated that eliminates the unwanted magnetic field and could also assist the sensor for proper functioning. It is observed that the most major issue striking the RLG takes place, when the Periscope prism fails to focus the outgoing wave onto the photodiode, which has certain external factors to be responsible of which magnetic field is the main factor. It is seen that sometimes due to external influences, gyro is affected by the magnetic field, so to de-magnetize the gyro, one method known as deguassing is used.

1.5 Commercially Available RLG Components (from G&H)

The RLG components are commercially availablr from many firms. These components from M/s G&H and their specifications are given below (Fig. 1.10a):

The RLG mirrors must have the following specifications:

$$\text{Reflectivity} = 99.995\%$$
$$\text{Scattering} : 1 - 3\,\text{ppm}$$
$$\text{Ansorption loss} : 1 - 2\,\text{ppm}$$

Very high damage threshold, temperature stability to have operating life of a few thousand hours (Fig. 1.10b).

Ring laser gyroscopes from G&H are deployed in commercial aircraft, missiles, satellites, and other military vehicles. The vertical integration used by them enables them to supply the entire ring laser gyroscope package: RLG frame and optical components. Their integration of capabilities delivers superior performance and streamlines sourcing. There are many components within the ring laser gryoscope (RLG)—each performs unique functions and is optimized accordingly. The Zerodur® frame must deliver maximal stability over a large range of operating environments—temperature, humidity, and environmental composition. G&H have 40 years of experience producing high-quality, consistent, Zerodur® frames. Zerodur has nearly zero expansion coefficient, which makes it ideal for making RLG blocks (Fig. 1.10c).

Flat, wedged, and curved mirrors used in RLGs are super-polished with surface roughness better than 1 Å RMS with high-reflectivity low-loss IBS coatings designed to survive extreme environmental conditions. Precision beam splitters, prisms, and wedges complete the assembly.

Extensive metrology capabilities are critical to the abilities of G&H to meet and exceed difficult specifications. They use the following instruments to achieve very stringent sprcifications of the RLG components:

(i) Zygo® ZeMapper™ for surface roughness and structure < 1 Å;
(ii) OGP® CMM for geometrical tolerances down to submicron features; and
(iii) 18" Zygo® Interferometer measures flatness up to 1/50th wave.

Other specifications are:
Bias stability:
1 degree/hour

Random walk: 0.1250 degree/(hour)1/2

Scale factor value: 116,000 ± 200 pulses per revolution

Operating temperature: −650 F to +1800 F

Shelf life: > 10 years

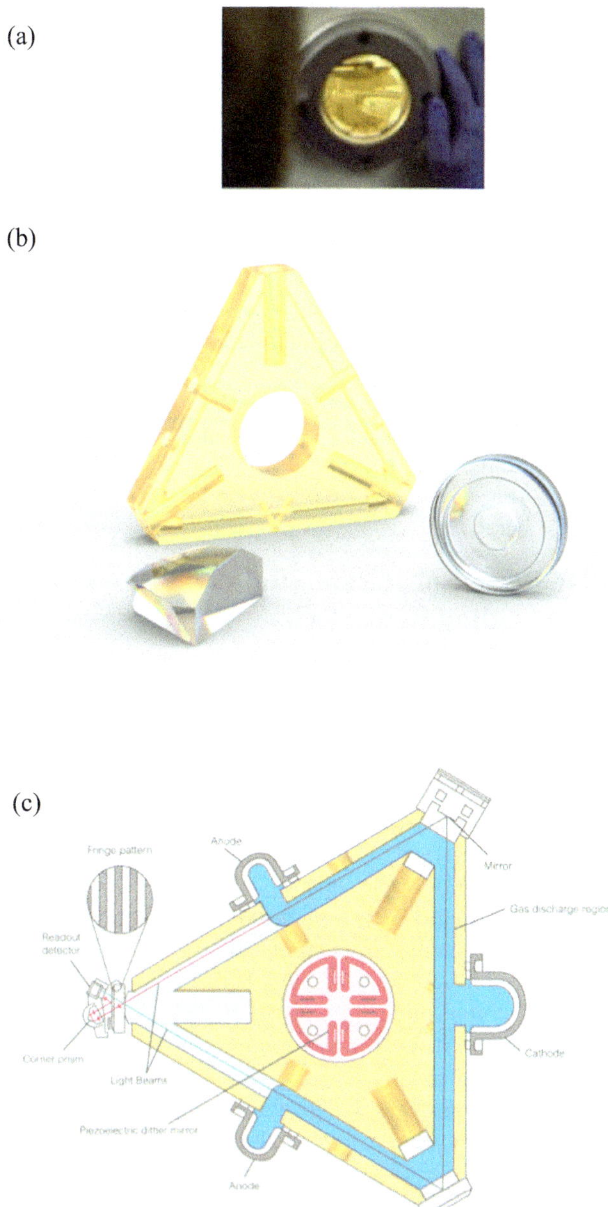

Fig. 1.10 **a** RLG mirror coating. **b** RLG components. **c** RLG block with the required attachments. Figure courtesy M/s G and H

Operating life: > or = 2000 h

Ring Laser Gyro (RLG) Mirrors Specifications.

Sub-Angstrom polished substrates, ring laser gyro (RLG) mirrors: RLG mirrors are made by using super-polished or laser quality polished substrates, since these are ideal for developing low-loss laser mirrors, in which thermal stability, surface scatter, and high laser damage thresholds are the critical requirements.

Some commercial firms including WZWOPTICAG use the state-of-the-art "super-polishing" technology to fabricate flat and curved substrates. In addition, this firm has the capability of measuring and verifying nearly all aspects of the mirror performance, mainly reflectance, scattering, surface figure, surface roughness, and absorption.

Super-polished surface of surface roughness of ≤1 Å rms, figured to ≤1/10 Wave PV, a scratch-dig of ≤10–5 and surface figure ≤/20 at 6328 Å, help in minimizing both scattering and absorption losses.

1.6 Qualitative Review of Recent Studies on RLGs and the Concluding Remarks

The recent applications of the ring laser gyroscope include an embedded Global Positioning Systems (GPS) capability for enhancing the accuracy of RLG Inertial Navigation Systems (INS)s on military aircraft, commercial airliners, ships, and spacecraft. Such hybrid INS/GPS units have been replacing their mechanical counterparts in most of the applications, though the spin gyro-based INSs are still used, where ultra-accuracy is needed.

For the fabrication of the RLGs, the super-polished Zerodur substrates are used, which are measured by phase-shift interferometer (PSI) and atomic force microscope (AFM). In general, it is observed that even though the substrate has very low surface roughness, there are some little bumps or hills, believed to due to the crystalline structure of the Zerodur composition. TIS of the mirrors is in the range 5–10 ppm after the annealing for about 1 h, and the absorption, which is ~100 ppm, is reduced to 40 ppm. The process of annealing leads to a decrease of the refractive index and extinction coefficient of the mirror material, though the changes are quite small. Recently, efforts [17, 18] have been made on the Inertial MEMS System Applications, and the Integrated Guidance Systems. More applications are in the fields of navigation, geophysics, relativity, symmetry testing, and quantum field theory. Li et al. [19] have suggested a novel calibration and compensation scheme in order to improve the precision of strap down inertial navigation system (SINS) and reduce the complexity of the traditional calibration method, by designing an optimization calibration method with four-direction rotations for calculating all error coefficients of RLG SINS in a series of constant temperatures.

Schreiber and Wells [20] have discussed the large ring laser gyroscopes, which are six orders of magnitude more sensitive than gyroscopes commercially available. It has been mentioned that partly, the increased sensitivity results from the scaled-up size, e.g., the largest of these gyroscopes, enclose an area of 834 square meters, which clearly implies that such RLGs are not compatible with navigation applications, and also, many corrections have to be applied, for taking into account a variety of factors, including the gravitational attraction of the moon. However, it has been emphasized that the progress in these devices has led to some novel applications in geodesy, geophysics, seismology, and testing theories in fundamental physics such as the effects of general relativity. Large RLGs are attached to the Earth's crust, so that a shift in the pattern, seen as an observed beat note in an actively lasing device, is directly proportional to the rotation rate of the Earth. Interestingly, any perturbations in this rotation rate capture the momentum exchange between the atmosphere, hydrosphere, and lithosphere, thereby making the large RLGs useful to indirectly monitor the combined effects of variations in global air and water currents, and also to supplement and improve calculations presently being made with very-long-baseline interferometry (VLBI) techniques, employed for measuring the orientation of the instantaneous rotation axis of the Earth and the length of day. Another important point to be noted is that any change in the ring's orientation shifts the beat note of the interferometer, making the large RLG useful for detecting tilts in the Earth's crust, which the current seismometers fail to distinguish from the horizontal acceleration.

Schreiber et al. [21] have demonstrated a 16 m^2 He–Ne RLG with sufficient sensitivity and stability to directly detect the Chandler wobble of the rotating Earth, and the detection of both the Chandler and the annual wobble has been verified by comparing the time series of the RLG measurements with the C04 series of Earth rotation data from the International Earth Rotation and Reference System Service. Yu et al. [22] have pointed out that no reported experimental work has dealt with the issue of characterizing the ARW of the constant rate biased RLG in the INS and have presented a cost-effective experimental approach to characterize the ARW of the gyros in the constant rate biased RLG INS.

Dell'Olio et al. [23] have discussed that the low-cost chip-scale optoelectronic gyroscopes having a resolution $\leq$10 °/h and a good reliability even in harsh environments have a strong impact on the medium-/high-performance gyro market, which is currently dominated by well-established bulk optical angular velocity sensors. The research and development activity aiming at the demonstration of such miniaturized sensors is crucial for aerospace/defense industry, and thus, it is attracting an increasing research effort and notably funds. Dell'Olio et al. [23] have reviewed the recent technological advances on the compact optoelectronic gyroscopes with low weight and high energy saving, by paying attention to both the so-called gyroscope-on-a-chip, which is a novel sensor, at the infantile stage, whose optical components are monolithically integrated on a single indium phosphide chip, and to a new ultra-high Q ring resonator for gyro applications with a configuration including a 1D photonic crystal in the resonant path. Also, the emerging field of the gyros based on passive ring cavities, which have already shown performance comparable with that of optical fiber gyros, has been discussed. Fan et al. [24] have proposed a dynamic burn-in grating model, in which the burn-in effect happens in the mirror not only in the standing wave state, but also in the traveling wave state, and it responds dynamically to beat frequency, which is different from the previous research results. It has also been emphasized that it can be derived that the lock-in threshold increases with the decrease of rotation rate for this model, and many puzzling phenomena of the RLG can be explained.

As is well known, the limitations to the performance of the RLGs come from the non-linear dynamics of the laser. Beghi et al. [25] have found a set of critical parameters affecting the time stability of the system, by following the Lamb semi-classical theory, and have proposed a method for estimating the long term drift of the laser parameters, and for filtering out the laser dynamics effects from the rotation measurement. It has been emphasized that the parameter estimation procedure, based on the perturbative solutions of the laser dynamics, makes it possible to apply Kalman filter theory for the estimation of the angular velocity. The Honeywell GG1320AN digital ring laser gyro [26], developed recently, is an affordable inertial sensor with the electronics and ring laser gyro packaged into a compact unit, which is simple in operation and use. With its digital I/O system, it is possible to integrate this gyro into most of the systems, and hence, it is used in many commercial, industrial, and military applications. By supplying an input voltage, it is possible to receive a digital output signal dependent on rotation. This gyro has all the advantages of the advanced technology, developed and used by Honeywell during the last four decades, and so is in the form of a reliable sensor for all types of inertial sensing. Its performance has been reported as: start-up time 1 s, bias stability less than 0.04 degree/hour, and angular random walk less than 0.04 degree/root-hour. It measures 45 mm high and is 87.6 mm in diameter, and weights 450 grams. Its temperature range is corresponding to an operating altitude of up to +21 km. Passaro et al. [27] have provided an overview of the current gyroscopes and their roles based on their applications, which presents the discussion of gyroscopes including mechanical gyroscopes and optical gyroscopes at macro-and microscale. Zhanshe et al. [28] have explained that the Microelectromechenical Systems (MEMS) gyroscope is widely used in many occasions to measure the angular speed of the moving objects and attracts the attentions of many research institutions all over the world because of various advantages of this type of sensor. Ma et al. [29] have critically reviewed the development and evaluation of passive optical ring resonator gyroscopes (OPRGs), by technically discussing the countermeasures against the parasitic noise sources, encountered in the OPRG including backscattering. Srivastava and Pattnaik [30], by using theoretical analysis and simulations, have proposed a gyroscope, which uses a ring resonator with a reflecting element. This structure is based on utilizing the rotation-induced Sagnac shift along with the inherent coupling between the clockwise and counterclockwise beams. Nazir et al. [31] have suggested the technique of temperature stabilization in fiber-optic gyroscopes for high-altitude aircraft. Srivastava et al. [32] have reported the first proof-of-principle demonstration of the resonant optical gyroscope with reflector earlier proposed by them. It has been emphasized that this device is very different from traditional optical gyroscopes, because it uses the inherent coupling between the clockwise and counterclockwise beams. Luo et al.

[33] have proposed a new virtual gyroscope with multigyroscope and accelerometer array (MGAA), for improving the performance of angular rate measurement. Interestingly, the outputs of the virtual gyroscope are obtained by merging the signals from gyroscopes and accelerometers. McAlpine [34] has developed a small, inexpensive, and highly accurate gyroscope, at the University of Michigan, which is expected to help drones and autonomous cars to be able to stay on track without a GPS signal. This gyroscope is claimed to be 10,000 times accurate and only 10 times expensive as compared to the gyroscopes used in the typical cell phones. McAlpine [34] has claimed 0.00016 deg/$\sqrt{}$hr angle random walk (ARW) and 0.0014 deg/hr bias instability (BI) from a 5.2 M-Q and 1-cm precision shell integrating (PSI) gyroscope. Schaijk et al. [35] have discussed that the external optical feedback (EOF), caused by the return loss of connected circuits, is well known to have a strong impact on the performance of semiconductor lasers, and these feedback effects are usually prevented by placing an optical isolator at the output of the laser. It has been emphasized that a strong isolation value of about 60 dB is required for tunable lasers. Hence, it has been established that it is really difficult to implement sufficiently strong optical isolator for photonic integration. Therefore, a ring laser with weak intracavity isolation is used to enable a feedback insensitive integrated tunable laser, in which the isolator is used only to enforce unidirectional operation of the device and the returned signal couples in the lossy counter-propagating direction. It has been shown that 10 dB intracavity isolation is able to completely suppress the effects of EOF over the operating range of interest in terms of output power, linewidth, and relative intensity noise, and in such a cavity, the weak isolation can be implemented with known integration schemes.

Badaoui et al. [36] have reported their recent progress toward a solid-state ring laser gyroscope (RLG), in which mode competition is circumvented by active control of differential losses, and nonlinear effects are mitigated by longitudinal vibration of the gain medium. It has been emphasized that the resulting dynamics is significantly different from that of a classical helium–neon RLG, mainly because of parametric resonances occuring when the Sagnac frequency is an integral multiple of the crystal vibration frequency. It is expected that the main experimental and theoretical results obtained so far will soon lead to the practical applications.

Wang et al. [37] have discussed that the ring laser gyro (RLG) dither axis bends and exhibits errors due to the specific forces acting on the instrument, which are known as g-sensitive misalignments of the gyros. The g-sensitive misalignments of the RLG triad result in severe attitude error in vibration or maneuver environments, in which case, large-amplitude specific forces and angular rates coexist. However, mostly g-sensitive misalignments are neglected while calibrating the strapdown inertial navigation system (SINS), and so, the results have small errors.

Wang et al. [37] have solved this problem and have proposed a novel method for calibrating the g-sensitive misalignments of an RLG triad in linear vibration environments. In this method, the SINS is attached to a linear vibration bench through outer rubber dampers, and so, the rocking of the SINS can occur when the linear vibration is performed on the SINS. Hence, the linear vibration environments can be created to simulate the harsh environment during aircraft flight. Thus, by analyzing the mathematical model of g-sensitive misalignments, the relationship between attitude errors and specific forces as well as angular rates is established, by which a calibration scheme with approximately optimal observations is designed. Finally, the vibration experiments are conducted to calibrate g-sensitive misalignments of the RLG triad. As expected, the vibration tests show that SINS velocity error decreases significantly after g-sensitive misalignments are compensated. This technique has been found to be very useful in strapdown inertial navigation systems.

Metadet dataset [38] has described that the essential elements for characterizing the performance of a laser gyro are (i) a bidirectional ring laser, (ii) a lightweight, efficient instrument, (iii) a high sensitivity to rotation,and (iv) a linear response without dead band. it has been realized that a high sensitivity to rotation has been addressed by substantially enhancing it through large dispersion d*n*/d*f*. The efforts are being made to experimentally demonstrate this enhancement, along with demonstrating the absence of dead band in a solid-state laser. An important point is the realization that it is possible to design and develope a mode-locked laser, in which the pulse envelope velocity is controlled by some parameters other than the dispersion.

Interestingly, this property has been demonstrated [38] in a mode-locked laser with intracavity Fabry-Perot and with intracavity resonant atomic vapor. Research efforts are being made to exploit this property by inserting a Fabry-Perot and a rubidium cell in a ring mode-locked Ti:sapphire laser, to demonstrate simultaneously (i) the enhancement of the gyro sensitivity, (ii) the use of a solid-state gain medium in a gyro, and (iii) the absence of dead band. More exciting experiments are being carried out to implement these results in a mode-locked fiber laser gyro, for demonstrating the light weight and efficient instrument required for space applications. In view of all these novel studies, it can be safely concluded that the subject is evolving and also progressing fast.

References

1. B. Lantz, R. Schofield, B. O'reilly, D.E. Clark, D. DeBra, Requirements for a ground rotation sensor to improve advanced LIGO. Bull. Seismol. Soc. Am. **99**(2B), 980–989 (2009)
2. A. Cygan, D. Lisak, P. Masłowski, K. Bielska, S. Wójtewicz, J. Domysławska, R.S. Trawiński, Pound-Drever-Hall-locked, frequency-stabilized cavity ring-down spectrometer. Rev. Sci. Instrum. **82**, 063107 (2011)
3. F. Zarinetchi, S. Ezekiel, Optics Lett. **11**, 6 (1986)
4. G.A. Sanders, M.G. Prentiss, S. Ezekial, Optics Lett. **6**, 11 (1981)
5. Sunada Satoshi, Tamura Shuichi, Inagaki Keizo, Harayama Takahisa, Ring-laser gyroscope without the lock-in phenomenon. Phys. Rev. A **78**, 053822 (2008)
6. R. Graham, Ring laser gain medium. M.Sc. Thesis, University of Canterbury, Christchurch, New Zealand, 2006
7. A.E. Siegman, *Lasers* (University Science Books, 1986)
8. C.P. Wyss, D.N. Wright, B.T. King, D.P. McLeod, S.J. Copper, G.E. Stedman, Collision broadening and quantum noise in a very large ring laser gyroscope. Opt. Commun. **174**, 181–189 (2000)

9. H. Yu, W. Wu, M. Wu, G. Feng, M. Hao, Systematic angle random walk estimation of the constant rate biased ring laser gyro. Sensors **13**(3), 2750–2762 (2013)
10. T. Qu, K. Yang, X. Han, S. Wu, Y. Huang, H. Luo, Design of a superluminal ring laser gyroscope using multilayer optical coatings with huge group delay. Scientific Reports 4 (2014) Article number: 7098
11. K.N. Chopra, Minimization of scattering loss of dielectric mirrors for ring laser gyroscope. Atti Fond G. Ronchi **70**, 179–187 (2015)
12. M. Salit, K. Salit, P.E. Bauhahn, Increasing the scale factor of a ring laser gyro via spectral hole burning. Proc. SPIE **8273**, 82730H (2012)
13. M. Salit, K. Salit, P. Bauhahn, Prospects for enhancement of ring laser gyroscopes using gaseous media. Opt. Express **19**, 25311 (2011)
14. J.E. Schaar, H.N. Yum, S.M. Shahriar, Theoretical description and design of a fast-light enhanced helium-neon ring-laser gyroscope. Proc. SPIE 7949, 794914 (2011)
15. T. Qu, K. Yang, X. Han, S. Wu, Y. Huang, H. Luo, Design of a superluminal ring laser gyroscope using multilayer optical coatings with huge group delay. Scientific Reports **4**, Article number: 7098 (2015)
16. C. Ashok, P.D. Lakshmi, K.C. Das, Design of optimal degaussing electronics for ring laser gyroscope. Int. J. Innovation Res. Electr. Electron. Instrum. Control Eng. **2**(9) (2014)
17. K.N. Chopra, Improvement in the laser induced damage threshold (LIDT) by the dual ion beam sputtering (DIBS) technology. Atti Fond G. Ronchi **70**, 395–406 (2015)
18. H.J. Cho, J.C. Lee, S.H. Lee, Design and development of an ultralow optical loss mirror coating for zerodur substrate. J. Opt. Soc. Korea **16**(1), 80–84 (2012)
19. J. Li, Y. Ma, X. Chen, Error modeling, calibration, and nonlinear interpolation compensation method of ring laser gyroscope inertial navigation system. Abstr. Appl. Anal. **2013**, 359675 (2013)
20. K.U. Schreiber, J.P. Wells, Invited review article: Large ring lasers for rotation sensing. Rev. Sci. Instrum. **84**(4), 041101 (2013)
21. K.U. Schreiber, T. Klügel, J.P. Wells, R.B. Hurst, A. Gebauer, How to detect the chandler and the annual wobble of the earth with a large ring laser gyroscope. Phy. Rev. Lett. **107**(17), 173904 (2011)
22. H. Yu, W. Wu, M. Wu, G. Feng, Hao Sensors (Basel) **13**, 2750–2762 (2013)
23. F. Dell'Olio, T. Tatoli, C. Ciminelli, M.N. Armenise, Recent advances in miniaturized optical gyroscopes. J. Europ. Opt. Soc. **9**, 14013 (2014)
24. Fan Zhen-Fang, Luo Hui, Lu Guang-Feng, and Hu Shao-Min, Research and discussion on the lock-in threshold variation of ring laser gyro. Acta Phys. Sin, 2012, 61(18): 184204
25. A. Beghi, J. Belfi, N. Beverini, B.A. Bouhadef, D. Cuccato, A. Di Virgilio, A. Ortolan, Compensation of the laser parameter fluctuations in large ring-laser gyros: a Kalman filter approach. Appl. Opt. **51**(31), 7518–7528 (2012)
26. Honeywell/My Aerspace, GG1320 Digital Ring Laser Gyro https://commerce.honeywell.com/webapp/wcs/stores/servlet/eSystemDisplay?catalogId=10251&storeId=10651&categoryId=53411&langId=-2. Honeywell International Inc. (2014)
27. V. Passaro, A. Cuccovillo, L. Vaiani, M. De Carlo, C.E. Campanella, Gyroscope technology and applications: a review in the industrial perspective. Sensors. **17**(10), 2284 (2017)
28. G. Zhanshe, C. Fucheng, L. Boyu, C. Le, L. Chao, S. Ke, Research development of silicon MEMS gyroscopes: a review. Microsyst. Technol. **21**(10), 2053–2066 (2015)
29. H. Ma, J. Zhang, L. Wang, Z. Jin, Development and evaluation of optical passive resonant gyroscopes. J. Lightwave … (2016). - ieeexplore.ieee.org
30. S. Srivastava, B. Pattnaik, Resonant optical gyroscope with 'reflector': analysis and simulations. J. Opt. **45**(2), 152–157 (2016)
31. J. Nazir, T. Vivek, T. Jaisingh, Temperature stabilization in fibre optic gyroscopes for high altitude aircraft. Optik. **127**(20), 9701–9710 (2016)
32. S. Srivastava, R.D. Shhreesha, H. Nandakumar, Novel optical gyroscope: proof of principle demonstration and future scope. Scientific Reports **6** (2016) Article No. 34634- nature.com

33. Z. Luo, C. Liu, S. Yu, S. Zhang, S. Liu, Design and analysis of a novel virtual gyroscope with multi-gyroscope and accelerometer arra. Rev. Sci. Instrum. **87**, 085003 (2016)
34. McAlpine Kate, *Small, Precise and Affordable Gyroscope for Navigating Without GPS*. (Khalil Najafi Group, University of Michigan, Tech Xplore, 2020)
35. T. Van Schaijk, D. Lenstra, K. Williams, E. Bente, Feedback insensitive external-cavity semiconductor ring laser utilizing a weak intracavity isolator, in *Photonic Integration, Research output: Chapter in Book/Report/Conference proceeding › Conference contribution › Academic, 2019 Conference on Lasers and Electro-Optics Europe and European Quantum Electronics Conference* (CLEO/Europe-EQEC, 2019)
36. N. El Badaoui, B. Morbieu, P. Martin, P. Rouchon, J.P. Pocholle, F. Gutty, G. Feugnet, S. Schwartz, Towards a solid-state ring laser gyroscope. C.R. Phys. **15**(10), 841–850 (2014)
37. W. Lin, W. Wu, G. Li, X. Pan, R. Yu, Ring laser gyro G-sensitive misalignment calibration in linear vibration environments. Sensors **18**(2), 601 (2018). https://doi.org/10.3390/s18020601
38. Metadata Updated: 2 May 2019, https://catalog.data.gov dataset

Chapter 2
Fiber-Optical Gyroscopes

2.1 Introduction

A fiber-optic gyroscope (FOG) is an optical device for sensing the changes in orientation, and thereby performing the function of a mechanical gyroscope, and for its operation is based on the interference of light having passed through a coil of optical fiber of very large length ~5 km. The advent of the diode lasers and low-loss single-mode optical fiber for the telecommunications industry led to the use of the Sagnac effect in the fiber-optic gyros (FOGs) in the form of the development of the practical devices in the beginning of this century [1–3]. Also termed as the interferometric fiber-optic gyroscope (IFOG), it is at present considered as an important option for various applications including the inertial navigation and guidance systems for aircraft, space industries, and helicopter attitude control. The success of this device has been due to its inherent advantages of solid-state technology, i.e., guided-wave optics and low-voltage low-power electronics, which have resulted in the cost reduction, and thereby enlarging the spectrum of its applications. IFOGs are based on the Sagnac effect, i.e., the production of a phase difference proportional to the dot product of the rotation rate vector by the area vector enclosed by the optical path, in a ring interferometer and, hence, has the advantage of single-mode optical fiber as the propagation medium. The FOG performance is affected by various critical system components and design characteristics, the most important of which are the coil optical fiber; the active source; the passive and integrated-optics components; and the detection systems.

It is now known that the ring laser gyroscope (RLG), at present, is well established in the medium and high-performance markets, since it has many advantages over mechanical gyros like: digital output linear with angular rotation, high sensitivity and stability, quick reaction times, and insensitivity to acceleration and immunity to most environmental effects. Still, the RLG is considered as a specialized instrument whose utility varies with the application, and several factors limit its selection over modern mechanical system, namely: The exacting cavity geometries and precision mirrors required for RLG construction and the necessity of assembly under stringent

K. N. Chopra, *Optoelectronic Gyroscopes*, Progress in Optical Science and Photonics 11,
https://doi.org/10.1007/978-981-15-8380-3_2

clean room conditions, driving its cost very high, bigger size, and larger weight of the RLG, as the solid glass optical block and mechanical dither assembly found in most RLGs unavoidably add to their weight. Because of these limitations, development of the fiber-optic gyros has picked up. Two main classes of fiber-optic gyros under development are: the interferometric fiber-optic gyro (IFOG) and the resonant fiber-optic gyro (RFOG). The RFOG appears to have better potential accuracy, but at present, it has the less mature technology. Interestingly, the RFOG devices resemble the RLGs and require a narrowband light source besides relying on an optical cavity, which is formed from the optical fiber tuned so that only one frequency of light propagates.

2.2 Interferometer Fiber-Optic Gyroscope

A very simple type of interferometer fiber-optic gyroscope is as shown in Fig. 2.1.

Its operation is based on the injection of two beams from a laser into the same fiber, but in the opposite directions. Because of the Sagnac effect, the beam traveling opposite to the rotation experiences a slightly shorter path delay than the other beam, and the resulting differential phase shift is measured through interferometry, by translating one component of the angular velocity into a shift of the interference pattern which is easily measured photometrically. Beam splitting optics is used for launching light from a laser diode into two counter-propagating waves (clockwise and anticlockwise directions) through a coil consisting of many turns of optical fiber. It is quite obvious that the strength of the Sagnac effect is dependent on the effective area of the closed optical path, which in fact is the geometric area of the loop, though enhanced largely by the number of turns in the coil.

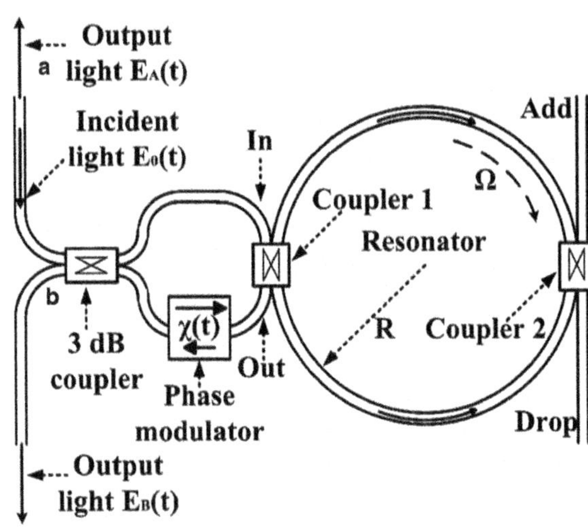

Fig. 2.1 Schematic of Add-drop ring resonator-based IFOG. Figure courtesy Zhang et al. [4]

2.2.1 Scale Factor

The scale factor of the FOG K is given by:

$$K = \frac{(4A)}{\lambda P}, \tag{2.1}$$

where A is the area bounded by the FOG, and P is the perimeter of the FOG, given by: $P = \pi D$, D being the diameter of the loop, AN is the total enclosed area. The interesting step taken is that the effective area is increased by winding N loops of the fiber-optic cable.

$$\Omega = \frac{\lambda c}{2LD}, \tag{2.2}$$

where L is the total length of the fiber, λ is the wavelength of light in the medium, Ω is the angular rate of rotation, $L = N\pi D$, and $\Delta\phi$ is the phase difference introduced, which is given by:

$$\Delta\phi = \frac{2\pi LD\Omega}{\lambda c}. \tag{2.3}$$

The FOG frequency difference $\Delta\phi$, termed as the beat frequency is proportional to the product of the geometric area enclosed by the light beams and the angular rate of cavity, and also inversely proportional to the product of the vacuum wavelength of the laser and the optical path of cavity, which implies the sources of scale factor error. Just to know the order of the values, the computations show that for a typical FOG (200 m-long fiber wounded on a 10 cm-diameter coil) the measurement of the Earth angular rotation $\Omega_e = 15°/\text{h} = 0.73\ \mu\text{r/s}$ requires to detect a phase difference $\Delta\phi = 36\ \mu\text{r}$, corresponding to an optical path difference of the order of 10^{-12} m. An important point to be noted here is that this measurement should be performed in DC, which is a much more difficult proposition than the measurement of a very small vibration in AC.

One other important parameter that the designer has to consider while designing the FOG of the required specifications is the phase noise at the quantum limit corresponding to a noise equivalent rotation rate $NE\,\Omega$, which is given by:

$$NE\,\Omega = \frac{\lambda c}{(8\pi AN)}\left[\frac{2hfB}{P_0\eta}\right], \tag{2.4}$$

where h is the Planck's constant, $f = c\lambda$, P_0 is the equivalent detected power, B is the measuring bandwidth, and η is the quantum efficiency of the photodetector. So the quantum limit corresponding to the noise equivalent rotation rate $NE\,\Omega$ has to be considered carefully for the required values of the measuring bandwidth, and quantum efficiency of the photodetector, and the three parameters have to be optimized.

The work on FOG started about four decades back, but because of its utility, the development of both the passive interferometer type of FOG (or IFOG), and a newer concept—the passive ring resonator FOG (or RFOG) has picked up recently both in the establishments and commercial firms. Its utility in inertial navigation systems is due to its various advantages: (i) A FOG is capable of providing extremely precise rotational rate information, because of its lack of cross-axis sensitivity to vibration, acceleration, and shock; (ii) unlike the classic spinning-mass gyroscope, the FOG has no moving parts and is free from relying on inertial resistance to movement. Being the most reliable alternative to the mechanical gyroscope, due their intrinsic reliability, FOGs are very suitable for high-performance space applications. Initially, the FOG generally showed a higher resolution than a ring laser gyroscope, but suffered from greater drift and worse scale factor performance. However, during the last two decades, the development of the FOGs has achieved large success and consequently greater reliability in the inertial navigation systems of many guided missiles, apart from being used in the surveying, stabilization systems, and in remotely operated vehicles and autonomous underwater vehicles, being implemented in both open-loop and closed-loop configurations. However, some of the disadvantages of FOGs are: (i) FOGs require calibration for determining the indication corresponding to zero angular velocity, as compared to the ring laser gyroscopes (RLGs) which do not require this since the zero beat frequency always means zero angular velocity.

Though, the ideal ring laser gyros produce pulses displaying an exact incremental change in angle, the accumulated gyro pulses are in general corrupted by both long-term and short-term errors. The long-term errors are due to the instability of gyro compensation parameters and include: (i) Bias error, (ii) scale factor error, and (iii) input axes misalignment error, which in the more demanding applications may be further decomposed into temperature-sensitive terms. The short-term errors cannot be calibrated and include: (i) dither spillover, (ii) random walk, and (iii) quantization noise. In fact, the random walk is the major limitation in reducing the required time of ground gyro compassing prior to system flight, since it produces an attitude error, which increases as a function of time. The random walk error is quite complex and includes four components—angle random walk (rate white noise) coefficient, bias instability coefficient, rate random walk coefficient, and ramp coefficient, which are determined by Allan variance method, which is a method of representing rms random drift error as a function of averaging time and is found very useful in the specification and estimation of random drift coefficients in a previously formulated model equation.

(a) Discussion of the Sources of Error

Fundamental limitations: (i) Sensitivity is limited by shot noise that goes as the square root of the power; (ii) the power received at the detector decreases with fiber length. However, it is obvious that Sagnac effect increases with the length of the fiber. The computations and experimental results reveal that for a sensitivity of 10^{-3}deg/h, L

is ~ few Km. Also, the output intensity I is modulated by the phase shift $\Delta\phi$ between the two beams, as:

$$I = I_o\{1 + \cos(\Delta\phi)\}. \tag{2.5}$$

It has to be noted that the dynamic range of the device can be easily configured by using the length and the diameter of the fiber loop. The calculations show that for

(i) wavelength $= 850$ nm, $L = 1$ km, $D = 10$ cm, $\Omega = 73\,\text{deg/s}$; and thus, for 1 m radian sensitivity, the minimum measurable value is = 0.084 deg/h; and for (ii) wavelength = 850 nm, $L = 100$ m, $D = 3$, the corresponding values are, respectively, 2400 deg/s and 2.8 deg/h. The designer has to optimize the values for achieving the particular sensitivity for a particular rate of rotation. This requires skill of the designers, who have to make the necessary corrections after the feedback from the obtained results.

In the elementary configuration of the IFOG, to reduce the parasitic noise, a low coherence light source, e.g., a semiconductor super-luminescent diode (SLD) is used. It has also to be noticed that the rotation rate from a very small phase difference, can be determined, only if both light beams travel in the fiber in exactly the same path and propagation mode, maintaining the reciprocity. It is clear that from the practical point of view, a single-mode fiber is required to be used. Also, it is noticed that even at rest, there is some difference between the two waves due to the different propagation delay for each polarization state, which results in the bias fluctuation. This problem is easily solved by using a polarizer, and thus ensuring the two waves to have exactly the same polarization. Interestingly, two couplers and two detectors have to be used in the system. Two couplers are required, because the transmission and coupling characteristics of a fiber coupler are different, and both waves must experience these two characteristics in the same manner. Consequently, the two detectors are now able to measure both waves in a "reciprocal" manner. The phase modulator is used to create a bias phase difference of $\left(\frac{\pi}{2}\right)$ between the two beams, for increasing the sensitivity of the measurement. It can be visualized that without this phase shift, the signal at the detector has zero gradient at $\Omega = 0$. In fact, the phase shift puts the signal in a position with high gradient at $\Omega = 0$, which provides a better measurement resolution, and in which case, the intensity at the detector in Eq. (2.3) is modified as given below:

$$I = I_o\{1 + \cos(\Delta\phi) - \left(\frac{\pi}{2}\right)\}. \tag{2.6}$$

In general, a phase difference of 1 μ rad is quite a measurable quantity, and therefore, μ can be defined as the rotation velocity which corresponds to the phase difference given as:

$$\Omega_\mu = \left\{\frac{\Omega_\pi}{(\pi.10^{-6})}\right\}. \tag{2.7}$$

(b) Discussion of the Sources of Noise

Fundamental limitations: (i) Sensitivity is limited by shot noise which is given as the square root of the power; (ii) the power received at the detector decreases with fiber length. However, in this case, the Sagnac effect increases with the length of the fiber. These two compensating effects of Sagnac effect determine the limit to the sensitivity. Interestingly, the time dependent temperature gradient along the length of the fiber may produce the spurious phase shifts because of the temperature dependence of the index of refraction and the cable, given by:

$$\Omega_{\mathrm{T}} = \frac{n_{\mathrm{c}} L^2 \Delta T}{24 N A} \left(\frac{\mathrm{d}n_{\mathrm{c}}}{\mathrm{d}T} + n_{\mathrm{c}} \alpha \right) \delta t, \tag{2.8}$$

where n_{c} is the refractive index of the cable material, T is the temperature, α is coefficient of the temperature dependence of the cable, and δt is the time duration. For an idea about the parameter dT, it may be noted that for dt = 1 h, D = 20 cm, L = 1.56 km, dn/dT = 10 − 5/°C, a = 5*10–7/°C, n = 1.45, and calculated shot noise limit of 0.0078 deg/h, dT needs to be 6.7×10^{-3}/°C. It has been reported that for better performance of the system, we must (i) use fibers with smaller dn/dT, (ii) wind the coil such that equidistant points from fiber center are physically close to each other.

Another important point to be noticed is that the single-mode fibers allow the transmission of two orthogonal polarizations. It is observed that the mechanical stresses can cause the power transfer between polarizations and birefringence, which leads to change in the wave velocity, which finally results in the super-position of two interference patterns that lead to poor contrast and fringe shifts. Polarization preserving fibers with high birefringence is used for suppressing the weak polarization. In practice, the polarization-maintaining fiber is used for the gyro coil, which takes into account both—polarization crosstalk and birefringence. Also, the FOG scale factor (the proportionality coefficient between the Sagnac phase shift and the rotation rate), is automatically calculated from the fiber length and the gyro coil diameter. The Sagnac component functions to model the Sagnac phase shift for both—the constant rotation rate and time-varying rotation rate. The user is able to simulate both phase and polarization non-reciprocity (PNR) by using the phase shift and polarization rotation models.

Backscatter at the output-input couplers is another problem, which can interfere with the main beams creating parasitic interferometers. The criterion followed in this case is that the reflected power should be of the order of the intrinsic Rayleigh scattering. For single-mode fiber (1 m long), the backscattered power is ~(–60 dB), and the reflected power at the input/output interface should be $< 10^{-6}$ times the beam power. It is important to note that though the antireflective coating on fiber surfaces solves this problem somewhat, it is not sufficient as a solution. The problem is solved by (i) using immersion cell to reduce index of refraction step; and (ii) production of fibers with slanted end faces.

Another noise factor is the uneven thermal fluctuations, which is one serious problem encountered in the making of the high-precision IFOGs, since the index of refraction of the optical fiber is a function of temperature, resulting in a light wave traveling throughout the fiber experiencing propagation delay due to a change in temperature along the fiber during its passage from one point at a certain time and a certain temperature, to another point at another time and another temperature, which is clearly very likely to be different for the co-rotating wave and the counter-rotating wave, for the case of the temperature distribution along the turned fiber being not symmetrical with respect to the center of the fiber. The two waves undergo different phase changes during their propagation along the optical fiber, which results in an output error. This problem can be reduced by: (i) choosing a fiber with a low temperature coefficient, and so a low dependence of the index of refraction on the temperature, (ii) developing unique winding technologies for ensuring that all portions of the turned fiber, located at symmetrical positions around its center, are positioned side by side, and (iii) monitoring the temperature along the fiber and making the compensating corrections for this at the FOG's output.

Optical Kerr effect is another parameter playing an important role in the performance of the system, which is attributed to the fact that the electric fields of the counter-propagating beams can cause changes in the index of refraction, which is non-reciprocal if $|E_1|$ and $|E_2|$ are not equal, and it has been observed and reported that $\frac{I_1}{I_2} = 10^{-4}$ leads to 10^{-3} deg /h shifts, which is really a serious source of error. Velocity of the clockwise and counter-clockwise waves is respectively given by the following equations:

$$v_{\text{CW}} = \frac{c}{[\varepsilon_o + \varepsilon_2(|E_1|^2 + 2|E_2|^2]^{1/2}}. \tag{2.9}$$

and

$$v_{\text{CCW}} = \frac{c}{[\varepsilon_o + \varepsilon_2(|E_2|^2 + 2|E_1|^2]^{1/2}}. \tag{2.10}$$

It may be noted that, for $d = 10$ cm, $A = 45\ \text{cm}^2$, wavelength $= 633$ nm, and $\vec{\Omega} = 15\,\text{deg/hr} = 7.3 * 10 - 5\,\text{rad/s}$, $\Delta\phi$ comes out to be $4.2^{*}10^{-8}$ rad. Thus, the designer has to be careful while choosing the value of $\Delta\phi$ for the values of d and A, for a workable FOG, required for measuring the particular rotation rate.

This problem is normally solved by employing square wave modulation. Some researchers have used the optical source represented by an SLD centered at 1531 nm, along with the multifunction integrated optical circuit (MIOC) including two-phase modulators in a push-pull configuration, driven by a square wave with a period equal to the fiber coil transit. It is clear that the resulting 90° phase shift biases the FOG around the point of maximum slope in the curve intensity versus rotation rate, maximizing its sensitivity, as shown in Fig. 2.2a.

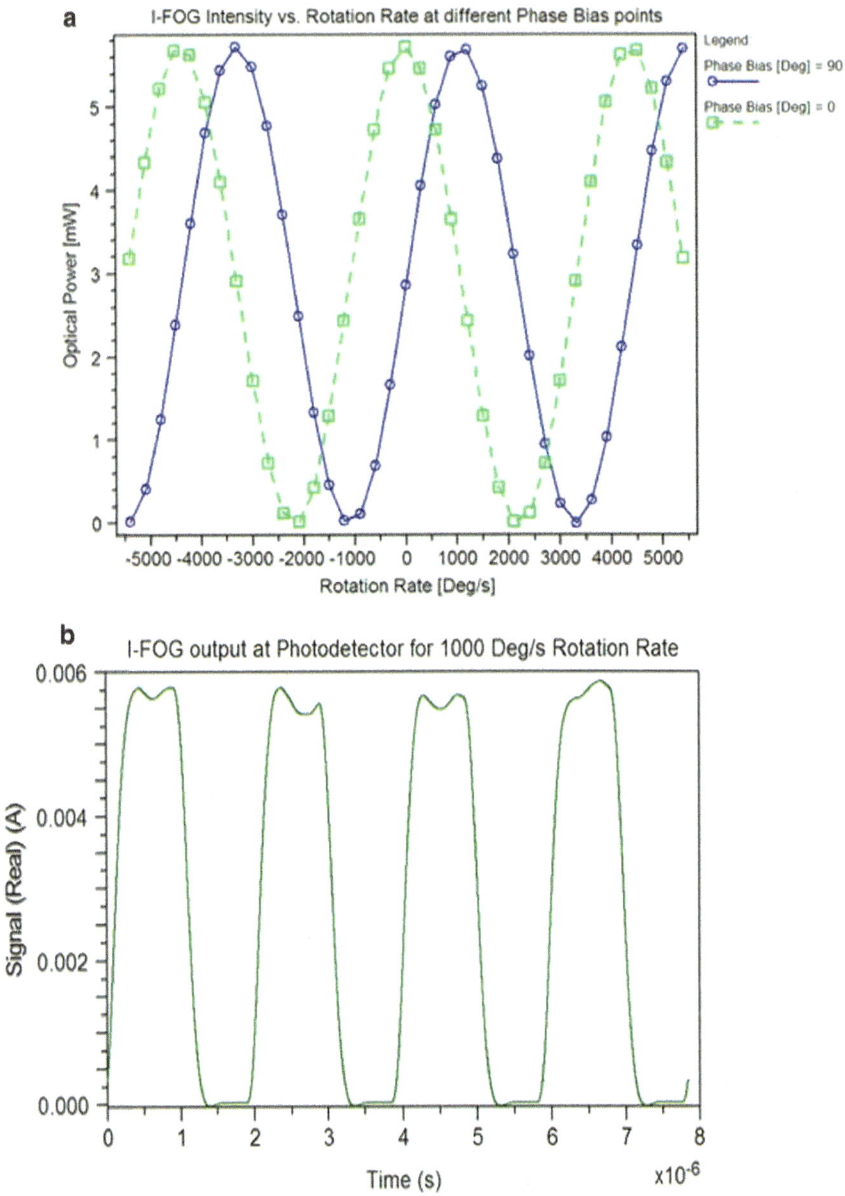

Fig. 2.2 **a** Maximizing the FOG sensitivity by biasing the FOG with 90° phase shift around the point of maximum slope in the intensity versus rotation rate curve, the two curves (left and right) are respectively for phase shift (degrees) = 0 and 90. **b** The real signal versus time 10^{-6} s curve for the IFOG output at the photodetector for 1 K Deg/s rotation rate. Figure courtesy optics.synopsys.com

In addition, the results of the real signal versus time 10^{-6} s curve for the IFOG output at the photodetector for 1 K Deg/S rotation rate, as reported in the literature have been reproduced here in Fig. 2.2b.

It is clear that the peaks are observed at the regular intervals (~0.2 ms in this case). Also, the peaks are quite well defined, with the maxima and minima having good contrast.

Zhang et al. [4] have proposed the modulation period and amplitude of the typical square wave phase bias modulation (SWPBM) applicable to a resonator-based interferometric fiber-optic gyroscope (R-IFOG) and have also theoretically studied the performance of the R-IFOG under SWPBM. It has been shown that (i) the R-IFOG possesses a performance distinct from that under the hypothetical time-independent phase bias, (ii) the sensitivity of the R-IFOG with SWPBM to a slow rotation rate is boosted in comparison with that without phase bias, and the rotation direction can be indicated, and (iii) the ultrahigh sensitivity can be attained by an R-IFOG of an extremely short fiber length when the R-IFOG with SWPBM consists of a high finesse resonator. It has been emphasized that the SWPBM of the proposed modulation period and amplitude enables highly sensitive and compact integrated closed-loop R-IFOGs.

The theoretical value of the sensitivity of the fiber-optic gyroscope is determined by the photon shot noise, emerging from the statistical distribution of the energy of the photons hitting the detector, and the formula describing the change in the measured angular velocity for the IFOG due to photon shot noise is given by:

$$\Delta\Omega = \left[\frac{\lambda c}{\{2LD(n_{\mathrm{p}})^{\frac{1}{2}}\eta_{\mathrm{D}}\tau\}}\right], \tag{2.11}$$

where n_{p} is the is the number photons hitting the detector per second, η_{D} is the detector. *s* quantum efficiency, and τ is the averaging time taken for the intensity measurement. Clearly, the number of photons n_{p} is related to the measured power P and is given by:

$$n_{\mathrm{p}} = \frac{P\lambda}{hc}, \tag{2.12}$$

where h is the Plank's constant. Just to have some idea about the power, and the phase difference, the computations by using Eq. (2.10), show that for a fiber coil of length $L = 1$ km, and diameter $D = 0.1$ m, with a wavelength of 850 nm; and assuming a quantum efficiency $\eta_D = 0.3$ and an averaging time of $\tau = 1$ s., typical power of $P = 1$ mW, which is equivalent to a number of $n_{\mathrm{p}} = 4.3 \times 10^{15}$ photons/s. The result is: $\Delta\Omega = 3.6 \times 10^{-8}$ rad/s, which is 0.0075 deg/h. Thus, the designer has to choose the values of L, D, η_{D}, and n_{p}, corresponding to a workable FOG, required for measuring the particular rotation rate. Though, some software is available, yet the experience and skill of the designers play very important role in optimizing the parameters.

Digonnet et al. [5] have reported a detailed experimental and theoretical analysis of a novel type of FOG, which utilizes an air-core photonic-bandgap fiber (PBF) in its sensing coil, the benefits of using the air-core fiber being the dramatic reduction of the phase bias drift due to temperature (Shupe effect), magnetic field (Faraday effect), and optical nonlinearity (Kerr effect), all of which are due to the fact that the fiber mode now propagates in air instead of silica. It has been stressed that the reduced Kerr sensitivity, combined with the low theoretical limit of backscattering in air-core fiber, presents the unprecedented potential of ultimately driving this type of FOG with a laser instead of a broadband source, which would yield lower noise and greater scale factor stability.

It has been established that a FOG is able to give extremely accurate rotational rate information, due to (i) being free from cross-axis sensitivity to vibration, acceleration, and shock; and (ii) not having any moving parts and thus not relying on inertial resistance to movement. The FOG has a higher resolution than a ring laser gyroscope. FOGs are designed in both open-loop and closed-loop configurations. FOGs, in general, require initial calibration, implying to find the indication corresponding to zero angular velocity. Also, some FOG designs are quite sensitive to vibrations. However, these systems can be made suitable for high shock environments, by coupling with multiple-axis FOG and accelerometers, and hybridizing with GNSS data, and thus mitigating the impact.

Another important use of FOGs is as a navigation aid in remotely operated vehicles and autonomous underwater vehicles. These are used on surface vessels and submarines as the main navigation systems. In addition, these are used on land defense vehicles and weapons for accurate pointing and positioning; and in surveying operations.

2.3 Depolarized Interferometric Fiber-Optic Gyroscope (IFOG)

Parez et al. [6] have made a study of a novel design of a depolarized interferometric fiber-optic gyroscope (IFOG), and, by means of computational simulation tools, presented a full analysis and design of an interferometric fiber-optic gyroscope (IFOG) prototype based on a closed-loop configuration with sinusoidal bias phase modulation. In addition, they have presented the complete design of the different blocks, optical and electronic, including some novelties as the sinusoidal bias phase modulation, and the use of an integrator for generating the serrodyne phase-modulation signal.

They have also given a detailed calculation of most parameter values, and the plots of the resulting signals obtained from simulation tools, which serve as a guide for the technologists engaged in this evolving field. Their design study is based on focusing in the use of a standard single-mode optical fiber, thereby allowing a cost competition to the commercial IFOG, though at the expense of reduced sensitivity. Their design consists of an IFOG model accomplishing tactical and industrial grade applications with sensitivity $\leq$0.055 °/h.

It is important to note that this design has two important characteristics: (i) an optical subsystem with advanced conception: depolarization of the optical wave by means of Lyot depolarizers, which allows the use of a sensing coil made by standard optical fiber, instead of polarization-maintaining fiber, and (ii) a novel and simple electronic design, which incorporates a linear analog integrator with reset in feedback chain, the integrator being able to generate a serrodyne voltage wave to apply to phase modulator (PM), thereby helping in obtaining the interferometric phase cancelation. This particular feedback design with sawtooth wave generated signal for a closed-loop configuration with sinusoidal bias phase modulation was used for first time in this field. Just to give some idea about the specifications of this model, its sensing coil consists of an 8 cm average diameter spool, which contains 300 m of standard single-mode optical fiber (SMF-28 type) realized by quadrupolar winding; and for the working wavelength of 1310 nm, the theoretically computed values of threshold sensitivity and dynamic range are 0.052°/h and 101.38 dB (from $\pm$ 1.164 $\times$ 10^{-5} °/s up to $\pm$ 78.19°/s), respectively, and the scale factor (SF) nonlinearity is 5.404% relative to the full scale.

2.4 Fiber-Optic Gyroscope-Based INS System

Singh et al. [7] have described an inertial navigation system (INS) for autonomous quadcopter application, based on fiber-optic gyroscopes. The useful characteristics of FOG like good performance in terms of weight, size, sensitivity, robustness, and resistance to physical forces; make it very efficient and extremely useful in strap down altitude navigational systems using heading reference. Singh et al. [7] have made a good effort to design and test run an INS-based on FOG, which tries to satisfy the generic positioning requirements for typical unmanned aircraft system (UAS) applications (UAS means an unmanned aircraft and the equipment to control it remotely). They have given laboratory evaluations of testing the response of the system based on several test cases, including errors due to the alignment and those induced by the sensor itself. It has been claimed that their simulation results suggest good performance of the test system developed and good reproducibility of results.

2.5 Commercially Available FOG Specifications

(i) Bias immunity to acceleration (gravity) and vibration

Acceleration does not create any rotation like signal in the optical sensing coil. Therefore, FOG output is naturally free of g, g2 components inherent in all mechanical gyros including MEMS. There is no vibration rectification error. It is capable of withstanding 1200 g shocks, and 125 °C temperature atmosphere.

Some features are:

(a) Low polarization x-talk (PM) and (ii) low PDL and PMD (SM)
(b) No dead zone or hysteresis.
In the open-loop FOG, the rotational signal is directly detected, and therefore, this linear signal conversion does not create any dead zone or hysteresis inherent in other gyro types, e.g., MEMS and closed-loop gyros.
(c) Instant response
The time delay in the sensing coil is <2 μs, which is light transmission time. Still, the practical limit (<0.1 ms) is due to the mechanical delay in the gyro frame, as the main frequency typically exceeds 2 kHz.
(d) Some data about fiber coil: wavelength as specified, e.g., 1550 nm, fiber length 50 m to 5 km, fiber diameter as specified, e.g., 250 micron, and coil inner diameter from 15 to 120 mm.

2.6 Erbium-Doped Fiber Amplifiers

Erbium-doped fiber amplifiers (EDFAs) are defined as fiber amplifiers based on erbium-doped optical fibers.

Erbium-doped fiber amplifiers have so far been found to be the most important fiber amplifiers for use in the long-range optical fiber communications, as they can efficiently amplify light in the 1.5-μm wavelength region, where telecom fibers have their minimum loss. The core element of a fiber amplifier is a piece of rare-earth-doped fiber, which provides laser amplification by stimulated emission, on being optically pumped with other light injected into the fiber. Recently, some efforts are being made to use them in FOGs, since the gain of Er-doped fiber amplifiers is maximum at 1550 nm, which is suitable and very useful for use in FOGs, since they can reduce loss. However, not much work seems to have been performed in this direction, and hence, for the designer of FOG, it is important to understand the intricacies of the designing aspects and efficient working of erbium-doped fiber amplifier (EDFA). The experimental setup and operation principle are discussed below:

Setup and Operating Principle of EDFA.

A simple setup of erbium-doped fiber amplifier is shown in Fig. 2.3.

The core of EDFA is the erbium-doped optical, fiber, which is a single- mode fiber. The working is simple in that the active fiber is generally pumped with light from two laser diodes (termed as bidirectional pumping), though in some cases, unidirectional pumping in the forward or backward direction (termed as co-directional and counter-directional pumping) is also commonly used. Mostly, the pump light, used has a wavelength around 980 nm or around 1450 nm, which excites the erbium ions (Er^{3+}),

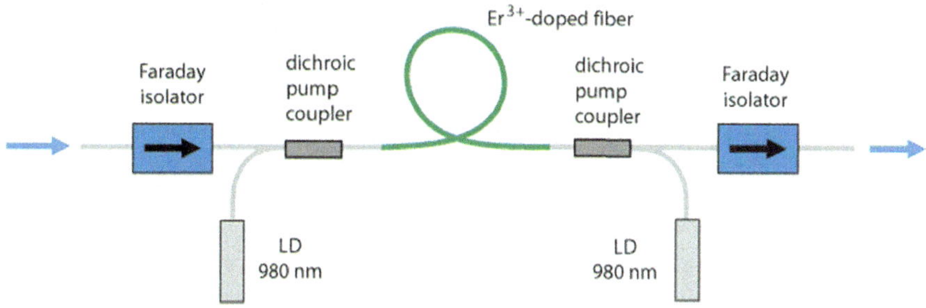

Fig. 2.3 Setup of a typical erbium-doped fiber amplifier. Figure courtesy Rüdiger Paschotta, RP Photonics Encyclopedia

respectively, into the 4I11/2 state and 413/2 state, from where they can amplify light in the 1.5-μm wavelength region by stimulated emission back to the ground-state manifold 4I15/2. The energy level structure of the trivalent erbium ion (Er^{3+}) and the optical transitions is shown in Fig. 2.4.

Two laser diodes (LDs) are used for pumping power for the erbium-doped fiber. The pump light is injected by dichroic fiber couplers. Pig-tailed optical isolators are used to reduce the sensitivity of the device to backreflections. The setup also contains two pig-tailed (fiber-coupled) optical isolators. The isolator at the input stops light coming from amplified spontaneous emission from disturbing any previous stages, and the other at the output suppresses or even stops lasing if output light is reflected

Fig. 2.4 Energy level structure of the trivalent erbium ion (Er^{3+}) and the optical transitions. Figure courtesy Rüdiger Paschotta, RP Photonics Encyclopedia

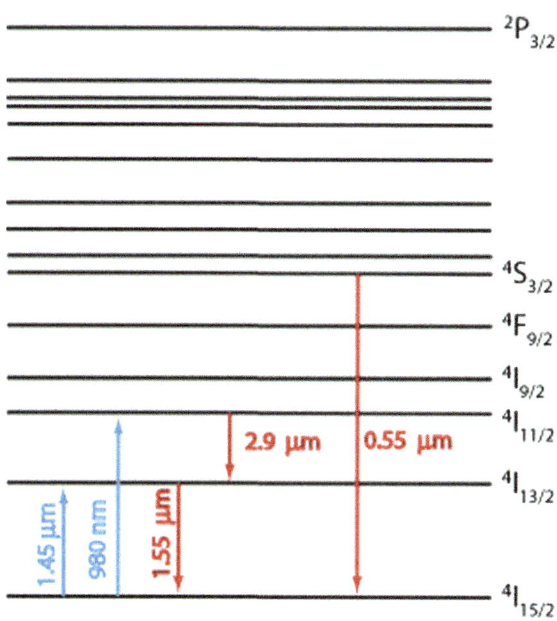

back to the amplifier. Thus, the isolators protect fiber amplifiers from being sensitive to backreflections. It is to be noted that if we are interested in getting a commercial fiber amplifier, we can order for having various other passive components apart from the optical isolators, to be contained in a commercial fiber amplifier. Some of the desired ones are (i) fiber couplers and photodetectors for monitoring optical power levels, (ii) pump laser diodes with control electronics, and (iii) gain-flattening filters. Even compact packages, with various passive optical components combined into a photonic integrated circuit, called planar lightwave circcircare available. Interestingly, very high signal gains, for the amplification of ultrashort pulses to high energies, can be obtained by using amplifier chains, formed by many amplifiers along with additional optical elements like isolators, filters, or modulators in between them. In view of this discussion, it is clearly understood that in order to use these fiber lasers for applications in FOGs, and the designer has to consider all these aspects and optimize the parameters. In addition, the designer has to consider other important parameters: (i) Shape of the gain spectrum, which depends on the absorption and emission cross sections, both in turn depending on the host glass; and (ii) the magnitude of the spectral shape of the gain, which is considerably influenced by the average degree of excitation of the erbium ions, since these have a quasi-three-level transition. A typical data curve for some variant of silica with additional dopants for avoiding clustering of erbium ions, as given in the literature has been shown in Fig. 2.5.

The final step in the process of light creation in a fiber laser is the binding of the spool of fiber by fiber Bragg gratings at each end, which are essentially mirrors for creating a high-power resonator on being placed at both ends of the doped fiber. The laser light resonates between these gratings, which reflect light at an angle depending

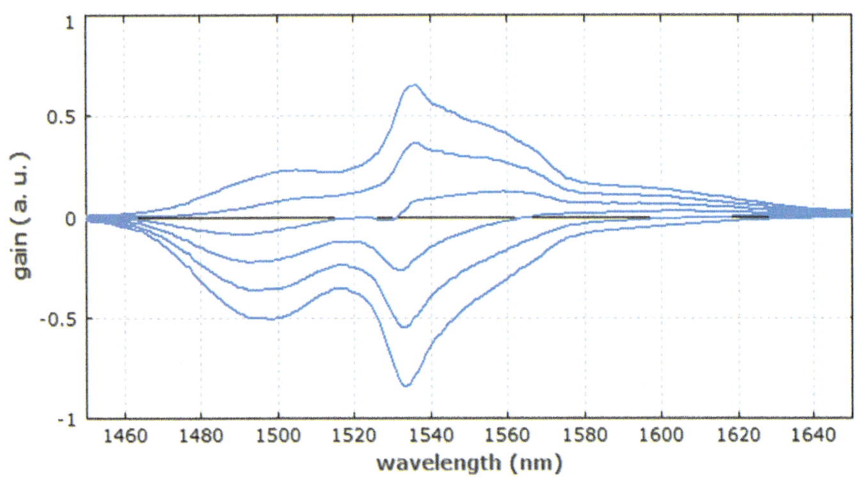

Fig. 2.5 Gain and absorption (negative gain) of erbium (Er^{3+}) ions in a phosphate glass for excitation levels from 0 to 100% in steps of 20%. Figure courtesy Rüdiger Paschotta, RP Photonics Encyclopedia

on the geometry of the reflecting surface. The light is bounced back and forth between the two gratings and gets amplified with every pass. The grating on the output of the fiber laser is partially transparent (Transmission less than 1%) for allowing the created longer wavelength laser light to exit the fiber.

2.6.1 Double Clad Fiber

Double clad fiber is an important class of optical fibers, which has a core doped with the rare earth. Though a standard single-mode fiber provides good beam quality characteristics, but its core requires the use of single-mode pump laser diodes. The single-mode core diameter has to be kept small to allow the single-mode laser oscillation, which yields the high-quality beam output. But the single-mode pumps needed to be used as pump sources do not have enough power to achieve the desired output power levels. Double clad fibers are able to overcome this problem by having a doped single-mode core with two layers of cladding around the core. The inner layer of cladding surrounding the core is pumped by high-power multimode pump laser diodes. The advantage is that the inner cladding can accept large amounts of pump light from multiple pump source coupled into the inner cladding layer. Thus, it is clear that the double clad approach provides optimization of high-power and high-beam quality.

A double clad fiber is shown in Fig. 2.6.

Above a threshold power level that depends on a number of parameters, and instabilities can cause a good single-mode beam to degrade into a poor-quality and unstable multimode signal.

Eidam et al. [8] have done an interesting study of the experimental observations of mode instabilities in a fiber laser (Fig. 2.7).

This transverse mode instability has been understood as the thermal effects occurring after power density in the fiber exceed a threshold value, below which the laser generates a stable single-mode beam of high quality. This effect is generally observed

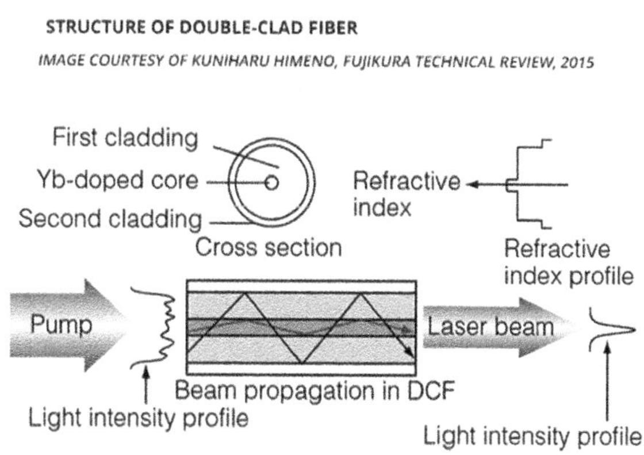

Fig. 2.6 Double clad fiber. Image courtesy of Kuniharu Himeno, Fujikura Technical Review No. 44, March 2015

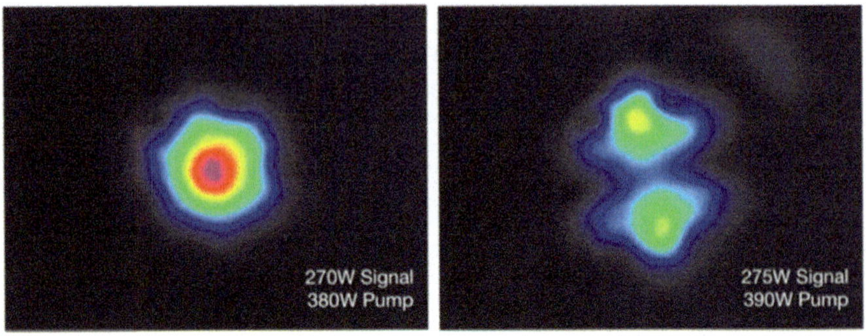

Fig. 2.7 Mode instabilities on a good single-mode beam quality. Figure courtesy Tino et al. [8]

at power levels of around a kilowatt, resulting in the beam quality to drop since the unstable higher-order transverse modes appear and, the laser starts operating in multi-mode pattern. Above this threshold power level, the laser output in the center of the beam varies quite chaotically. This level depends on many factors, like the structure of the active fiber, the seed power used in the final amplifier stage, and details of the system design. These effects are observed in both pulsed and continuous output fiber lasers. These transverse mode instabilities occur only for milliseconds and cause fundamental-mode power to vary from nearly zero to about 90% of the power, the rest of the power being transferred to the LP11x and LP11y modes. Interestingly, the interference between modes effectively forms a refractive index grating in the fiber, resulting in the instabilities.

2.7 Fiber Lasers

Fiber lasers have many advantages over other conventional lasers, as these are compact and rugged, and do not go out of alignment, besides easily dissipating thermal energy. They are of various forms, of technology similar to that of other types of lasers, but still having their unique advantages. The fiber laser is a solid-state laser, though quite different from other standard types of solid-state lasers, in the sense that the medium in this case, is a clad fiber in place of a rod, a slab, or a disk. Another big difference is that in fiber laser, light is emitted by a dopant in the central core of the fiber, and not a general medium, the core structure ranging from simple to quite complex. An important characteristic of the fiber lasers is that the fiber has a large surface-to-volume ratio to facilitate the heat dissipation. Another characteristic of their operation is that they are optically pumped, mostly with laser diodes, but in some cases with other fiber lasers. They are fabricated mostly by using the optics in the form of fiber components, fiber coupled to one another. However, some designs use bulk optics or even a combination of an internal fiber-coupling system with external bulk optics. There are many types of diode pumping sources: a single diode, an array, or many separate pump diodes, each with a fiber connected to a coupler. The doped fiber has a cavity mirror on each end for multiple reflections. Some designs use Bragg gratings, which can be fabricated within the fiber for compactness and ruggedness. The geometrical structure of the fiber used in fiber lasers is important, the commonly used geometry being a dual-core structure. An undoped outer core in the form of an inner cladding collects the pump light and guides it in the fiber. Thus, the stimulated emission generated in the fiber passes through the single-mode inner core. The dopant (Erbium or Ytterbium) present in inner core is stimulated to emit radiation by the pump light. The structure of a fiber laser is shown in Fig. 2.8a.

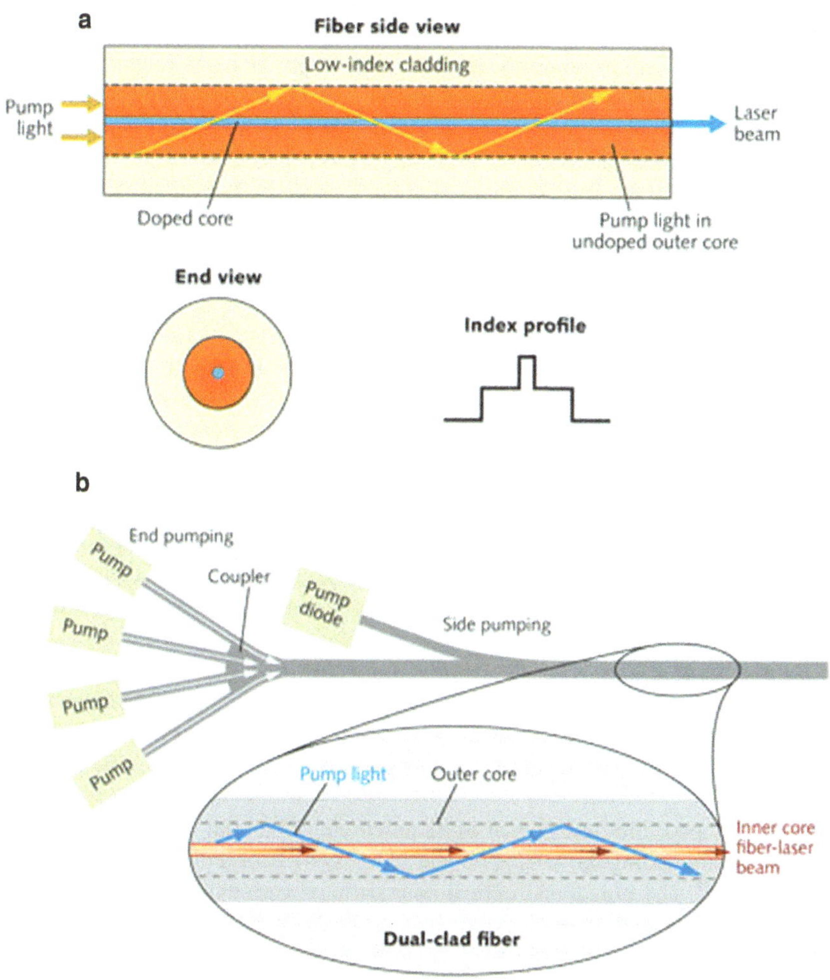

Fig. 2.8 **a** Structure of a fiber laser. **b** Fiber laser with end pumping. Figure courtesy Fiber Lasers: The state of the art, Laser Focus World April 1, 2012

The structure of a fiber laser contains a doped inner core (the laser), an undoped outer core (an inner cladding) through which the pump light is channeled, and an outer cladding. A fiber laser can be end- or side-pumped. In end pumping, the light from one or many pump lasers is fired into the end of the fiber. In side-pumping, pump light is coupled into the side of the fiber. Actually, it is fed into a coupler that couples it into the outer core. This is different from side-pumping a laser rod, where the light comes in orthogonally to the axis. A fiber laser with end pumping is shown in Fig. 2.8b.

Another type of fiber laser geometry is in which it is end-pumped by one or more lasers, or side-pumped by side-coupling pump light into the outer core. The designer has to consider many parameters for its successful and optimum working.

Coupling the pump light into the core, matching it to the optical absorption, and coupling the pump light into the inner core to produce a population inversion resulting in stimulated emission in the inner core. The laser core is observed to have various degrees of gain, depending on the doping in the fiber and the fiber length. The designer has to adjust and optimize these factors to get the desired performance. He has also to consider the power limitations arising particularly from working within a single-mode fiber, which has a very small cross-sectional area, and therefore, a very high-intensity light going through it, when the nonlinear Brillouin scattering becomes increasingly important, limiting the output at multikilowatt levels. The designer has to consider all these parameters, which requires great skill and experience.

2.7.1 *Difference Between a Fiber Laser and a Fiber Amplifier*

The main difference between a fiber laser and a fiber amplifier is just the cavity. Whereas, a fiber laser is a gain medium, a pump and a cavity, an amplifier has no cavity. Since the cavity allows to select the oscillating mode, there is no need to seed with a signal to generate a single frequency, as it is initiated by noise. However, an amplifier needs a signal seed for its operation.

An audio amplifier is just its optical counterpart, as it needs an input signal to amplify it. Therefore, if a microphone is brought closer to the speaker plugged to the amplifier, it forms a loop to produce a really annoying noise at a selected frequency.

For fiber laser, the cavity can be of various types, e.g., a ring cavity, a Bragg cavity, or mirrors.

Erbium-doped fiber has proved to be an efficient material system having many applications, including broadband optical sources, wideband optical amplifiers, and tunable lasers. The broadband optical sources have various application areas like optical device characterization, gyroscopes, and optical coherence tomography. The amplified spontaneous emission (ASE) in erbium-doped fiber has been found useful to construct light sources with the advantages of high output power and broad optical bandwidth. The erbium-doped fiber amplifier (EDFA), which was the first successful optical amplifier, has revolutionized the optical communication industry about three decades back, and presently are widely used in all kinds of fiber communication systems, especially proving very suitable for wavelength division multiplexed (WDM) systems.

The amplification window of an optical amplifier is the range of optical wavelengths for which the amplifier provides a usable gain, the amplification window being determined mainly by three parameters: (i) spectroscopic properties of the dopant ions, (ii) the glass structure of the optical fiber, and (iii) the wavelength and power of the pump laser.

It is surprising to note that though the electronic transitions of an isolated ion are very well defined, broadening of the energy levels occurs when the ions are incorporated into the glass of the optical fiber, which results in broadening of the amplification window. Interestingly, this broadening is of two types: homogeneous, in which all ions exhibit the same broadened spectrum; and inhomogeneous, in which different ions in different glass locations exhibit different spectra. The homogeneous broadening is caused by the interactions with phonons of the glass, while the inhomogeneous broadening occurs due to the differences in the glass sites, where different ions are hosted. In fact, the different sites expose ions to different local electric fields, which shift the energy levels by the Stark effect. Also, the Stark effect removes the degeneracy of energy states having the same total angular momentum, specified by the quantum number J. This explains as to why the gain spectrum of the EDFA has several peaks that are smeared by these broadening mechanisms; resulting in a very broad spectrum, e.g., 30 nm in silica. This characteristic of broad gain-bandwidth of fiber amplifiers makes them very useful in wavelength division multiplexed communications systems as a single amplifier can be used to amplify all signals being carried on a fiber, whose wavelengths lie within the gain window.

2.7.2 Difference Between the Advantages of a Fiber Laser and a Fiber Amplifier

Fiber lasers have many advantages including:

(i) No moving parts or mirrors in connection with the light-generating source, which makes it a much simpler design than gas CO_2 or disk lasers. Being free from the need for many maintenance checks and repairs, they help in saving time and money.
(ii) Higher electrical efficiency, because of which a 3 kW fiber laser is able to use one-third of the energy that a 4 kW CO_2 machine uses. Higher electrical efficiency results in quicker project times and less costs.
(iii) Great with reflective materials, which imply that the design of the fiber laser reduces the risk of backreflections ruining the machine, which is a common problem with other lasers. In fact, a fiber laser is able to perform easily on many materials like copper, gold, brass, and aluminum.

(iv) Faster cutting capabilities, which means that when using thinner materials, the fiber laser is able to cut much faster than other lasers, and with the same accuracy. To give some idea, the 3 kW fiber laser cuts a 1 mm stainless steel piece three times faster than a 4 kW CO_2 laser.

Erbium-doped fiber has proved to be an efficient material system having many applications, including broadband optical sources, wideband optical amplifiers, and tunable lasers. The broadband optical sources have various application areas like optical device characterization, gyroscopes, and optical coherence tomography. The amplified spontaneous emission (ASE) in erbium-doped fiber has been found useful to construct light sources with the advantages of high output power and broad optical bandwidth. The erbium-doped fiber amplifier (EDFA), which was the first successful optical amplifier, has revolutionized the optical communication industry about three decades back, and presently are widely used in all kinds of fiber communication systems, especially proving very suitable for wavelength division multiplexed (WDM) systems.

Fiber amplifiers have many advantages including:

(i) The first signal pulse is comparatively very strong, because it is amplified after a longer time of pumping the fiber. However, a high-peak power in a long fiber can generate a high Raman gain, which can cause detrimental effects of stimulated Raman scattering (SRS); which has to be taken care of by the design engineer of the system.
(ii) Er-doped fiber amplifiers have revolutionized the under sea transmission systems due to their outstanding characteristics and adaptation possibilities for further increasing the total throughput capacity from the first-generation single-channel systems for the wideband systems allowing amplification of about 100 channels. More work is in progress for the ultimate utilization of their potential bandwidth with seamless ultrawideband EDFAs designed for long-haul submarine applications.
(iii) It has been observed that the pulsed fiber amplifier systems are an interesting alternative to Q-switched bulk lasers, since they allow for having many interesting features. However, the problem is that the development of such a system is comparatively more complicated, as the designer has to deal with many issues, including gain saturation, ASE, and fiber nonlinearities. This requires thorough investigations based on numerical simulations for succeeding in developing highly efficient system with optimum performance.

2.8 Choosing Between Ring Laser Gyroscope and Fiber-Optic Gyroscope

As explained earlier, RLG (Fig. 2.9) and FOG (Fig. 2.10) are optical gyroscopes, both based on the Sagnac effect. Choosing between the two is a difficult proposition, as it depends on various factors. One difference is that the ring laser gyroscope has

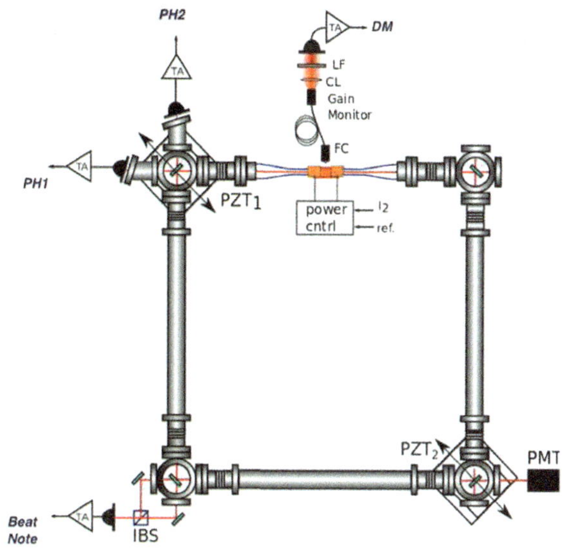

Fig. 2.9 Typical scheme of RLG with a square ring cavity. Figure courtesy, Di Virgilio et al. [29]

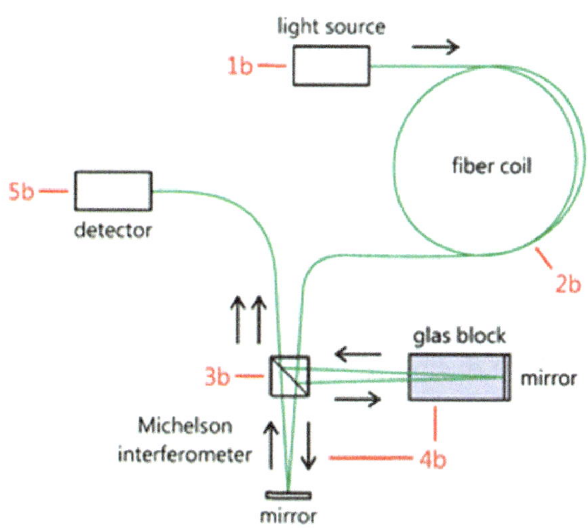

Fig. 2.10 Commonly used model of FOG

higher precision. In case of FOG, the various optical effects that actually occur make this technique more difficult to achieve (Fig. 2.9).

Whereas the fiber-optic gyroscope measures the phase difference of the light in two directions, the ring laser gyroscope measures light frequency difference in two directions. In case of FOG, the phase difference of light in two directions is measured through the *Y* waveguide device; in which, the phase difference/rotational speed coefficient can be increased, by enclosing a larger area, for which a longer optical fiber is required. Though the optical fiber has a lower cost, it has drawbacks of being susceptible to non-uniform thermal expansion and contraction due to temperature change, and tension variation during winding.

Another point to be considered is that though in case of RLG, the smaller surrounding area can have a relative frequency difference; still, the ring laser gyro measures and expresses the rotational speed by fringe count and does not require closed-loop control, since in this case, light travels in resonant cavity and is not much affected by outside, which increases its precision. However, the resonant cavity is much more difficult to make, besides being relatively more expensive.

Finally, if we have to choose, FOG is the best choice as compared with mechanical gyro or RLG, due to various factors including: Fewer parts, resulting in being firm and stable, and with strong resistance to both shock and anti-acceleration; larger length of the wound fiber, and so providing the detection sensitivity and resolution several orders of magnitude higher than those of the ring laser gyro; no mechanical transmission parts, resulting in no wear and tear, and hence a longer service life; easy to adopt integrated optical circuit technology, giving the stable signal, thereby making it suitable for digital output, and connecting with computer interface.

Some more points to be considered while making the choice are as given below:

RLG is heavier and bigger than FOG, about two times by weight, and more expensive than FOG. Also, the FOG is two times more precise than RLG. In addition, the power output of the RLG reduces with passage of time, and so the RLG becomes unusable after certain period of time. Thus, it is clear that in general, the designer of the system has to choose FOG instead of RLG.

Some of the desirable advantages in case of IFOG are as given below:

(i) High rate capability >1000 deg/s; (ii) no lock-in problem; (iii) immune to both radio frequency and electromagnetic interferences (EMIs); (iv) negligible g and g^2 sensitivity; (v) no moving parts and gas, thereby providing high operational reliability and long shelf life; (vi) fast activation time; and (vii) being photon-based, FOG is immune to space radiation, which affects electrons and the devices based on electronics.

However, the choice is not easy, as RLG has also a number of advantages as listed below:

(i) It gives digital output with angle increment; (ii) it has very high sensitivity; (iii) it has a stable input axis, and easy detection over a wide range of measurement (from 10 to 3"/h to thousands of "*i* s); (iv) it is insensitive to accelerations as it has no moving parts, thereby nullifying the chances of errors due to acceleration; and (v) it is very reliable.

Thus, it is quite difficult to choose between the two; and the system engineer has to carefully weigh the pros and cons of the two before making the final choice.

One more important point to be mentioned here is that some successful efforts have also been made to fabricate very efficient RLGs by using optically polished totally internally reflecting prisms in place of dielectric coated laser mirrors, the one used for output with minutely different angle. As even the state of the art double ion beam sputtered coatings result in marginally increasing the microroughness of the sample, and so the scattering from the mirrors; this technique is capable of marking more efficient RLGs. However, this approach can be followed, only if the designer has the ability of making optically polished prisms of very high quality, besides having art of getting highest degree of optical alignment. In general, this technique is suitable for those designers who are better in optical polishing than in depositing dielectric coatings.

2.9 Commercially Available FOG

A good performance FOG is now easily available by many firms. The fiber-optic gyroscope consists of an optical portion and a signal processing portion. A common structure of the FOG's optical portion using the integrated optical waveguide is as shown in Fig. 2.11a.

Light source selection and polarization control are the important components of the FOG. The light source gives the optical signal required for the fiber-optic gyroscope. The accuracy of the fiber-optic gyroscope depends mainly on the stability of each of the three parameters: optical wavelength, power, and polarization. The mostly used source by the commercial firms is the super-luminescent diode (SLD) or amplified spontaneous emission (ASE) by using erbium-doped fiber. The SLD sources have poor average wavelength stability and are therefore used for low precision fiber-optic gyroscopes. For FOGs required for high-precision applications, the ASE source is used, since it has a much higher spectral width and spectral stability than the SLD source, and the output power is also high. The ASE light source amplifies the spontaneous emission signal generated by the erbium-doped fiber by the

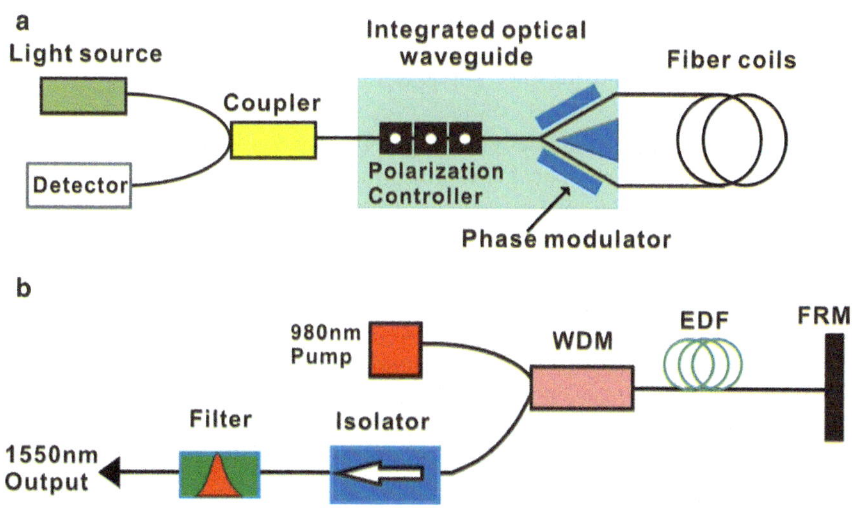

Fig. 2.11 **a** Schematic of the optical path of an interferometric fiber-optic gyroscope. **b** Specific structure diagram of the ASE source. Figure courtesy DK Photonics

action of pump light. The ASE light source uses the double-pass backward, and its specific structure is as shown in Fig. 2.11b.

As shown, 980 nm pump laser, wavelength division multiplexer (WDM): 980/1550 nm wavelength division multiplexer, Er-doped fiber (EDF), fiber rotation mirror (FRM), optical isolator, and optical filter are used in the setup. Polarization fluctuation is the other factor affecting the performance of the FOG.

The drift caused by polarization instability in the fiber-optic gyroscope is usually controlled by the fiber coil, which has the characteristic of adopting a polarization-maintaining fiber, so that the device in the optical path part also needs to adopt a polarization device, such as a polarization-maintaining coupler, and a polarization-maintaining isolator, in the integrated optical path. It is worth mentioning here that depolarization technology can also be used, in which the fiber coil requires only an ordinary single-mode fiber, and the device in the optical path part does not have to use the polarization-maintaining device, so that the polarization control device needs to use the fiber depolarizer.

2.10 Qualitative Review of Some Novel Studies on FOGs and the Concluding Remarks

At present, the fiber-optic gyroscopes (FOGs) represent an important development in the field of inertial sensors and are now considered an alternative technology to mechanical and ring laser gyroscopes for inertial navigation and control applications. Nayak [9] has presented the specifications, system configurations, and error and sensitivity analysis for different types of FOGs, including critical technology, and has addressed the design and production of a different FOG that fulfills requirements such as small size, low production cost, low-power consumption, and a broad spectrum of applications.

The simplest type of the fiber-optic ring resonator gyroscope (RFOG) consists of a recirculating passive optical cavity, in which the cavity resonant frequency is modulated by the angular rotation, and the change in the resonance frequency of the counter-propagating waves is transformed into a variation of the output power of the fiber-optic coupler. Obviously, this configuration operates as a high finesse (F) fiber ring resonator, and therefore, good performances may be achieved with a much shorter fiber length of a factor ~ F. Many novel types of the FOGs have recently been developed, and the stimulated Brillouin scattering has been investigated for an active fiber ring resonator, which exhibits in principle the capability of high performances mainly due to the reduced negative effects induced by non-reciprocity and non-ideal behavior of the employed components. Another type—the fiber ring laser gyro (FRLG) is based on the rare-earth-doped fibers as the active medium sustaining the laser oscillation in the sensing coil, configuration is similar being quite similar to the RFO.

It is now well established that the present high-grade IFOGs provide sensitivities of up to 0.01 deg/h, and the ring laser gyros [10], though having higher sensitivity, are complex and quite expensive. Though a lot of studies have been made on the RLGs [10, 11], which have been very useful in the inertial investigation systems, it has been found that making of the RLG mirrors with very low scattering loss (~10 ppm) is quite cumbersome, and needs state of art dual ion beam sputtering units and other innovations in the designing of the mirrors [12–14]. Some development work has started on the resonant fiber-optic gyroscope (RFOG), which is a passive resonator and has advantages of both that can be optimized according to the requirement. This is in fact a passive ring cavity fabricated by using an optical fiber in place of the active cavity used in the laser gyro, which eliminates the problem of mode locking in laser gyros. Also, the rate of change of the resonance frequency is quite similar to that in case of the laser gyro. At present, the development of a low-cost, solid-state, fiber-optic gyroscope meeting the main requirements of the navigation systems is drawing the attention of the research groups and the commercial firms.

Ciminelli et al. [15] have done an interesting study on the high performance in ring resonator for new generation monolithically integrated optical gyroscopes. Optical gyroscopes with high sensitivity are important rotation sensors for inertial navigation systems. Zhang et al. [16] have presented the concept of integrated resonant optical gyroscope constructed by active long-range surface plasmon polariton (LRSPP) waveguide resonator, in which the LRSPP waveguide doped gain medium is pumped to compensate the propagation loss, which has lower pump noise than that of conventional optical waveguide. It is reported that the peculiar properties of single polarization of LRSPP waveguide have been found to significantly reduce the polarization error. It has been emphasized that the metal layer of LRSPP waveguide is electro-optical multiplexed for suppression of reciprocal noises, which shows a limited sensitivity of ~10^{-4} deg/h, and a maximum zero drift which is four orders of magnitude lower than that constructed by conventional single-mode waveguide. Bolognini et al. [17] have presented a giant-area fiber-optic gyroscope (GAFOG) combining a Raman distributed temperature sensor (RDTS) for metrological monitoring of Earth rotation rate fluctuations, which exhibits sensitivity limits in Earth rotation sensing associated to temperature variations along the fiber. The advantage of this approach is that the distributed temperature measurements provide an accurate fiber temperature mapping, which enables the tracking of the signal phase drifts, which in turn allows the reduction of the phase noise at low Fourier frequencies, and a corresponding enhancement of the Earth rotation rate variation sensitivity beyond the currently achieved 2 nrad/s for integration times of 1000 s.

Recently, some innovative work has been undertaken to use a photonic crystal as a medium to develop a compact optical gyroscopes [18]. It has been well recognized that the photonics has a great potential to fabricate optical components which can be integrated with standard semiconductor technology. An interesting mechanism called coupled-cavity waveguide (a linear array of equally spaced identical local defects) has now been realized for light guiding inside the photonic crystal. Some more novel studies related to the use of the optical fibers in sensors and other fields have been undertaken. Song et al. [19] have reported the first measurement of the polarization crosscoupling of a 900-m length of quadrupolar-wound sensing coil for the fiber-optic gyroscope (FOG) using polarization-maintaining photonic crystal fibers (PM-PCFs). The crosscoupling has been analyzed with the optical coherence domain polarimeter (OCDP), and the polarization extinction ratio (PER) of the PM-PCF coil has been measured. Also, for the sake of comparison, a standard panda fiber coil with nearly the same length has been examined. It has been reported that their results show that PER of the PM-PCF coil is ~10 dB larger than the panda fiber coil, and polarization crosscoupling in the PM-PCF coil is ~10 dB less than in the panda fiber coil. Interestingly, based on measurements, a fourfold reduction in phase error has been demonstrated numerically in a FOG made of the PM-PCF coil, compared to the same gyro operated with a similar coil of panda fiber. It has been emphasized that these findings have important benefits in a higher measurement precision and stability of FOG. Yu et al. [20] have introduced a new design for a bidirectional optical sub-assembly for fiber-optic gyroscope applications that integrates a super-luminescent light-emitting diode, a photodiode, a beam splitter, an isolator, a fiber receptacle, and a thermal electric cooler, which is less than 1.5 cm in diameter and 2.5 cm in height. It has been reported that as the chip temperature is kept at 30 °C under environment temperature of −35 °C, 25 °C, and 75 °C, the bidirectional optical sub-assembly reaches stability at a center wavelength of 1,539 nm and a wavelength shift of 1.5 nm. A 3D simple model with the finite-element method has been used to analyze the thermal performance. Annamdas [21] has reviewed the developments in fiber-optical sensors and applications. It has been reported that (i) during the last two decades a steep rise in extensive research of optoelectronic and fiber-optical communication fields has been witnessed, which has resulted in applications focused initially in military, aerospace equipments, and later in health monitoring for medicine, heritage culture, and various engineering fields; and (ii) health monitoring has attracted the effort of many engineers throughout the world and is fast emerging as a pioneering field, which is helping to improve world economy in two ways; first, directly by providing longevity to all the important structures and secondly, by prevention of untimely failure leading to loss of life and money. Annamdas [21] has presented a complete over view of fiber sensing systems and their usefulness and has highlighted some of the general fiber optics available in the market with selected examples. Poudyal et al. [22] have done the simulations and theoretical analysis and have presented an innovative readout technique that uses the change in a beat frequency signal for sensing rotation in the resonant fiber gyroscope with a "reflector". Wang et al. [23] have demonstrated a novel dual-polarization interferometric fiber-optic gyroscope (IFOG), which requires only one coupler and no polarizer. This is an important development, since polarization non-reciprocity (PN) errors in conventional IFOGs increase quite significantly if the polarizer is absent, or if only one coupler is used. Lefèvre [24] has discussed in detail the challenges for FOGs to be one ultimate rotation sensing technology. Fan et al. [25] have systematically investigated the scale factor correction characteristics of two ring laser gyroscopes of different sizes. They have (i) provided the numerical figures for the variation in the scale factor, which can be 144 or 70 ppm for square gyroscopes with arm lengths of 8.4 cm or 15.6 cm, respectively, during frequency tuning and

(ii) have observed a dip in the scale factor at the line center of the gain characteristic for both gyroscope sizes. Another important observation made is that on exciting a different longitudinal mode, the scale factor behavior remains the same, but the scale factor values differ slightly from those computed by geometric considerations. It has been stated that the scale factor tends to decrease with increasing discharge current, and the sensitivity of the scale factor to variations in the excitation decreases with increasing discharge current.

Yao et al. [26] have discussed in detail the optimization of the FOG fiber coil; static parameters (the static parameters determine FOG performance at a fixed temperature, e.g., insertion loss, IL, and polarization extinction ratio—PER); and transient parameters (the temperature and vibration determine the transient properties of a FOG system, e.g., the asymptotic angular error, called the pointing error of a gyro system incorporating a quadrupolar fiber coil, which is linearly proportional to the final temperature gradient between the inner and outer coil surfaces in the radial direction, the pointing error thermal sensitivity, PETS being the proportionality constant).

Skalský et al. [27] have presented a simple method for compensating the Sagnac phase shift in an interferometric fiber-optic gyroscope (IFOG) with a piezoelectric modulator. The importance of this topic has been realized; and therefore, a special issue [28] is being planned to be published, which will provide a forum for the latest research activities in the field of integrated microphotonic and fiber-optic gyroscopes and their applications. It can be observed that in addition to the novel studies on the FOGs, research efforts are also going on the development of the compact optical gyroscopes, and therefore, it can be concluded that the subject is evolving fast.

References

1. S. Donati, G. Giuliani, M. Sorel, Proposal of a new approach to the electrooptical gyroscope: the GaAlAs integrated ring laser. Alta Freq. Riv. Elettron. **9**(6) (1997); also: Alta Freq. Riv. Elettron. **10**(6), 45–8 (1998); Integrated semiconductor laser rotation sensors, Proc. Photonics West, (San Diego, 1999) Integrated Optics Devices III, SPIE **3620**, 322–331
2. V. Lodi-Annovazzi, S. Merlo, Mechanical-thermal noise in micromachined gyros. Microel. J. **30**, 1227–1230(1999); also: R. Voss, Proc SPIE **3224**, 62–73
3. S. Merlo, M. Norgia, S. Donati, *Fiber gyroscope principles* (Handbook of Fibre Optic Sensing Technology. Wiley, London, 2000), pp. 1–23
4. X. Zhang, Y. Zhang, Y. Gai, Y. Wang, X. Liu, K. Wang, P. Yuan, Performance of a resonatot-based interferometric fiber-optic gyroscope under the square wave phase bias modulation. Opt. Eng. **53**, 102711 (2014)
5. M. Digonnet, S. Blin, H.K. Kim, V. Dangui, G. Kino, Sensitivity and stability of an air-core fibre-optic gyroscope. Meas. Sci. Technol. **18**, 3089 (2007)
6. R.J. Pérez, I. Álvarez, J.M. Enguita, Theoretical design of a depolarized interferometric fiber-optic gyroscope (IFOG) on SMF-28 single-mode standard optical fiber based on closed-loop sinusoidal phase modulation with serrodyne feedback phase modulation using simulation tools for tactical and industrial grade applications. Sensors **16**(5), 604 (2016). https://doi.org/10.3390/s16050604
7. B. Singh, H. Rana, S. Kumar, P. Bhulania, G. Minocha, A novel design of fiber optic gyroscope based INS system for U AS applications, in *3rd International Conference on Recent Trends in Computing 2015 (ICRTC- 2015)*, Procedia Comput. Sci. **57**, 1317–1323, (2015)

8. Tino Eidam, Christian Wirth, Cesar Jauregui, Fabian Stutzki, Florian Jansen, Hans-Jürgen Otto, Oliver Schmidt, Thomas Schreiber, Jens Limpert, Andreas Tünnermann, Experimental observations of the threshold-like onset of mode instabilities in high-power fiber amplifiers. Opt. Express **19**, 13218 (2011)
9. J. Nayak, Fiber-optic gyroscopes: from design to production. Appl. Opt. **50**(25), E152–E161 (2011)
10. K.N. Chopra, Analytical treatment of the optimization and mathematical modeling of ring laser gyroscopes and a short qualitative review. Atti Fond G. Ronchi **70**, 639–656 (2015)
11. C.P. Wyss, D.N. Wright, B.T. King, D.P. McLeod, S.J. Copper, G.E. Stedman, Collision broadening and quantum noise in a very large ring laser gyroscope. Opt. Commun. **174**, 181–189 (2000)
12. K.N. Chopra, Minimization of scattering loss of dielectric mirrors for ring laser gyroscope. Atti Fond G. Ronchi **70**, 179–187 (2015)
13. H.-J. Cho, J.-C. Lee, S.-H. Lee, Design and development of an ultralow optical loss mirror coating for zerodur substrate. J. Opt Soc. Korea **16**, 80–84 (2012)
14. K.N. Chopra, Improvement in the laser induced damage threshold (LIDT) by the dual ion beam sputtering (DIBS) technology. Atti Fond G. Ronchi **70**, 395–406 (2015)
15. C. Ciminelli, F. Dell'Olio, M.N. Armenise, F.M. Soares, W. Passenberg, High performance InP ring resonator for new generation monolithically integrated optical gyroscopes. Opt. Express **21**, 556–564 (2013)
16. T. Zhang, G. Qian, Y.Y. Wang, X.J. Xue, F. Shan, R.Z. Li, J.Y. Wu, X.Y. Zhang, Integrated optical gyroscope using active Long-range surface plasmon-polariton waveguide resonator. Sci. Rep. **24**(4), 3855 (2014)
17. G. Bolognini, C. Clivati, G. Costanzo, M. Frittelli, F. Levi, A. Mura, S. Schiller, D. Calonico, Giant-area fiber optic gyroscope with Raman-nased temperature sensing for metrological applications, in *Proceedings of SPIE 9157, 23rd International Conference on Optical Fibre Sensors*, 915724 (2 June 2014). https://doi.org/10.1117/12.2059316
18. B.Z. Steinberg, Rotating photonic crystals: a medium for compact optical gyroscopes. Phys. Rev. E **71**, 056621 (2005)
19. N. Song, P. Ma, J. Jin, J. Song, Reduced phase error of a fiber optic gyroscope using a polarization maintaining photonic crystal fiber. Opt. Fiber Technol. **18**(4), 186–189 (2012)
20. C.W. Yu, C.L. Chiu, T.T. Shih, J.J. Jou, A stable and compact optical module for fiber-optic gyroscope application. Fiber Integ. Opt. **33**(4), 306–314 (2014)
21. K.K. Annamdas, V.G. Annamdas, Review on developments in fiber optical sensors and applications. Int. J. Mater. Eng. **1**, 1–16 (2011)
22. S. Poudyal, S. Srivastava, R. Gowrishankar, Self-beating resonant optical gyroscope with a "reflector": the possibility of high sensitivities at reduced costs. Appl. Opt. **58**(7), 1699–1706 (2019). osapublishing.org
23. Z. Wang, Y. Yang, P. Lu, R. Luo, Y. Li, D. Zhao, Dual-polarization interferometric fiber-optic gyroscope with an ultra-simple configuration. Opt. Lett. **39**(8), 2463–2466 (2014). osapublishing.org
24. H.C. Lefèvre, Optical fiber technology, in *The Fiber-Optic Gyroscope: Challenges to Become the Ultimate Rotation-Sensing Technology* (Elsevier, 2013)
25. Z. Fan, G. Lu, S. Hu, Z. Wang, H. Luo, Scale factor characteristics of laser gyroscopes of different sizes. Rev. Sci. Instrum. **87**, 045002 (2016). http://doi.org/10.1063/1.4945348, Published Online: 11 Apr 2016 Accepted Mar 2016
26. S. Yao, S. Wey, W. Daab, Characterization of both the static and dynamic parameters of fiber coils ensures defect-free and consistent manufacturing for reliable fiber-optic gyroscope (FOG) devices. Fiber-Opt. Sens. Transf. FOG Coil Manuf. Art Sci. Laser Focus World (2017)
27. M. Skalský, Z. Havránek, J. Fialka, Efficient modulation and processing method for closed-loop fiber optic gyroscope with piezoelectric modulator. Sensors **19**(7), 1710 (2019). https://doi.org/10.3390/s19071710
28. Special Issue, Recent Advances in Integrated Microphotonic and Fiber-Optic Gyroscopes", A special issue of Sensors (ISSN 1424-8220). This special issue belongs to the section "Optical Sensors". Recent Advances in Integrated Microphotonic and Fiber-Optic Gyroscopes

29. A.D. Di Virgilio, N. Beverini, G. Carelli, D. Ciampini, F. Fuso, E. Maccioni, Analysis of ring laser gyroscopes including laser dynamics. Special Article—Tools Exper. Theo. Euro. Phy. J. C **79**, 573

Chapter 3
Minimization of Scattering Loss of Dielectric Mirrors

3.1 Introduction

The dielectric mirrors for ring laser gyro (RLG) are different from the conventional laser mirrors used in various laser systems, mainly on the basis of very low scattering loss (5–10 ppm), required for avoiding the lock-in problem encountered in the RLG, because of the scattering from the mirrors, resulting in the fringes becoming stationary, in spite of the rotation of the system, and thereby changing the optical path difference between the two laser beams traveling in the opposite directions, which leads to the failure of the sensor, and hence stoppage of the feedback for applying any midcourse corrections to the vehicle on which it is mounted. As discussed in the first chapter, the RLG consists of a ring laser having two counter-propagating modes over the same path in order to detect rotation and is based on the principle of the Sagnac effect which shifts the nulls of the internal standing wave pattern in response to the angular rotation. The RLG coatings (RLGCs) can be fabricated by considering various parameters—(i) clean environment of ~100 class in the cleaning room and the coating room, especially near the coating chamber; and the use of the vacuum pumps not using any oil; (ii) the substrate should be made of the material (glass) of the laser class of very high quality, having a flatness ~$\lambda/5$, very low scratch and dig of a number ~3 each, (which means that the maximum scratch width is 0.003 mm, maximum dig or bubble diameter is 0.003 mm, and dig or bubble separation distance is 1 mm), and microroughness ~5–10 A°, so that the scattering remains very low even before coating; (iii) the coating materials, which should be pure and must have absorption edges away from the wavelengths of interference, along with having other characteristics like—compatibility with each other and having good adhesion, minimum strain, and high mechanical strength; (iv) the polishing technique used for polishing the substrate—the usual polishing techniques of abrasives can give a smooth surface of very good quality and surface figure, but not sufficient for developing the RLG coatings. Any abrasive material left in the substrate at some scratch or dig becomes a center of absorption and scattering. These techniques are able to provide the surface of rms value of roughness ~15 A°, and not

K. N. Chopra, *Optoelectronic Gyroscopes*, Progress in Optical Science and Photonics 11,
https://doi.org/10.1007/978-981-15-8380-3_3

lower, which becomes the source of scattering and absorption. The float polishing technique (a commonly used float polishing technique is a non-contact polishing tool utilizing a magnetic compound fluid) for optical substrates is used to overcome this problem, for providing a substrate with the surface of rms value of roughness ~3 A°, and hence leading to considerable reduction in the scattering loss of the coatings; (v) the coating design (the optimum pair design), which should produce shifting of electric field maximum away from the boundaries of the multilayer stack, i.e., interfaces between the different layers, which are the centers of the highest scattering. The achievement of the minimum amount of scattering for the multilayer dielectric coatings for laser systems depends on thin film design, which for the case of the minimum number of layers, minimizes the absorption and scattering from the layers. This design requires the materials for the high- and low-refractive indices to be such that their difference is maximized, which results in minimizing the number of layers for achieving the same degree of reflection; and (vi) in addition, the coating technique is used, which should be capable of providing high packing density (PD) to the coatings, leading to the minimization of the voids in the films through which water and chemicals may diffuse resulting in changing the optical thickness of the coatings and also reducing the mechanical strength of the coatings and, therefore, affecting their stability of spectral characteristics.

The clean room facilities and the types of pumps are really very important for the successful fabrication of the RLG coatings. Clean room conditions are required because they minimize the risk of recontamination after cleaning the substrates and also while loading them in the coating chamber. Mostly, lint-free wipes without silicone constituent are used for cleaning manually in the final cleaning process, and the commonly used solvents are—methanol, isopropanol, or acetone. The problem of contamination from walls is completely avoided, by putting a lining of the aluminum foil or by having coaters using removable steel liners.

The DIBS technique [1] has a great advantage that there is no direct exposure of substrate to plasma, and so the intense bombardment by electrons and ions from the plasma is totally avoided. In view of the fact that, the deposition is carried out at a low pressure ($<10^{-2}$ Pa), the contamination of the growing films and the gas incorporation is minimized. Another great advantage of this technique is that the deposition parameters, e.g., ion energy and the angle of incidence at the target are precisely controllable and independent of each other. This technique produces the stable optical properties due to the formation of the non-columnar structure and results in: (i) better stability in the adverse environments like—high temperature and presence of reactive gases; (ii) higher PD, and so higher refractive index and hence higher stability with time; (iii) Lower grain size and so reduction in scattering; (iv) better adhesion, because of comparatively much larger energy of the atoms as compared to the other techniques; (v) reduction in stress and increase in mechanical strength; and (vi) better stochiometry due to the completeness of the reaction.

3.2 Optimization of the Process Control Parameters

The coating process is controlled by many parameters like—the rate of deposition, substrate temperature, oxygen partial pressure in case of the designs including dielectric metal oxides, thickness calibration, and the material-melt preconditioning. The optimization of these process control parameters is very important, as it determines the stoichiometry of the growing film, and helps in producing a homogeneous layer with the desired metal-oxygen content and structure. The materials like TiO_2 and SiO_2 have a tendency to break into lower oxides, and so the flow rate of O_2 has to be optimized while the coating of such materials. In addition to all these parameters, deposition of all the RLGCs, are done by giving planetary rotation (~30 rpm) to the substrate, for giving uniform scattering loss, and also by annealing of the coatings (at 300 °C), which has also been observed to increase the adhesion and mechanical strength of the coatings considerably (~15%). The optimum flow rate of O_2 in case of TiO_2 and SiO_2 has, respectively, been observed to be ~4 A°/sec and 8 A°/sec; and the optimum values of O_2 flow during evaporation have, respectively, been found to be ~10 cc.sec and ~15 cc.sec, for producing the optimum oxygen partial pressure. All these values are for the case of the EB evaporation. In case of the DIBS technology, the rates of deposition are very slow indeed (~1 A°/sec) for TiO_2, SiO_2, and Ta_2O_5 (which is quite commonly used in this technique).

3.2.1 The Process of the Formation of the Films

In conventional coatings like the electron beam (EB) evaporation, the condensing atoms have low energy (0.1–0.5 eV), and so low mobility, and as a result on reaching the substrate, are mostly deposited on or around the point of contact, leading to the formation of the columnar structure with low packing density (PD) and porosity, and hence the coatings have a lot of scattering from the surface (~250 ppm). The solution for reducing the scattering is provided by various methods: (i) a low rates of deposition; (ii) ion-assisted deposition (IAD); (iii) magnetron sputtering; (iv) ion beam sputtering, and (v) double ion beam sputtering. In the first case, the rates of evaporation by EB technique are reduced to ~3–5 A°/sec, which leads to the considerable reduction of the scattering. In the IAD, one ion beam gun is used to assist in the evaporation by imparting more energy to the atoms and is more effective than the first one. The magnetron sputtering is also quite effective, whereas the ion beam sputtering is much more effective. However, the DIBS technique, in which one beam gun is used for sputtering, and the second gun is used for assisting in coating and is extremely effective and is capable of producing the state of the art coatings. The minimization of the scattering loss of the coatings developed by the DIBS as compared to those by the conventional coating techniques has been explained by the flowchart given below (Fig. 3.1).

Conventional Coatings: Low Energy of Condensing Atom (0.1-0.5eV) >>>Low Mobility>>> Low P.D >> > Porous Films >> > Scattering Loss

Solution-DIBS - One Monoenergetic +ve Ion Beam Sputters Target Other cleans the Substrate and also assists in coating >>> High Energy of Condensing Atom (100- 500 ev >>> High mobility >>> High P.D. >>> Minimization of Scattering Loss

Fig. 3.1 Flowchart of the minimization of scattering loss of the coatings developed by the DIBS as compared to those by the conventional coating techniques

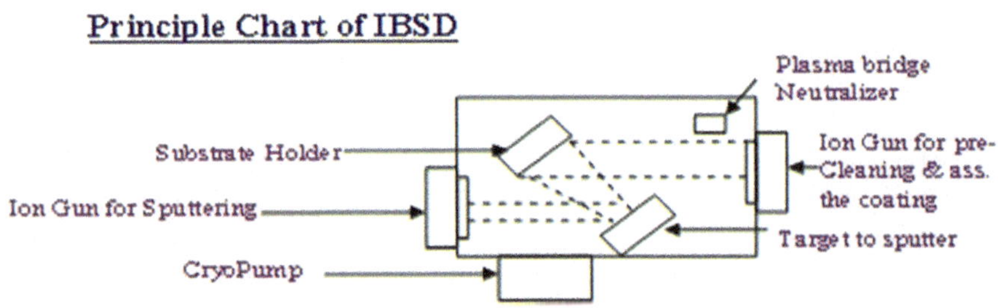

Fig. 3.2 Principle of DIBS deposition chamber

The process of reducing the PD and the principle of the ion beam sputtered deposition (IBSD) has been explained in the following figure.

Solution-IBSD

- Mechanical Disruption of the Columns by Beam of Energetic Ions [Several Hundred ev.]—Pushing Material Deeper into Columns and Squeezing out voids.
- Increased Reactivity of Ionized Gas—less absorption (Fig. 3.2).

3.3 Optimum Pair Design for Dielectric Mirrors

Designing of a multilayer stack for the laser mirrors is in the form of a quarter wave stack consisting of the layer sequences as given below:

$$[\text{Air}][HL]^n\text{Glass} \tag{3.1}$$

where H denotes a high index layer of quarter wave optical thickness (QWOT), L denotes a low index layer of quarter wave optical thickness (QWOT), and [HL] denotes the basic period, which is repeated n times. The reflectivity increases with increasing n and becomes very high indeed (~99.9%) for $n = 15$. It is important to note that the value of n required for a particular reflectance is minimized by maximizing the difference in H and L. Also, the lesser the number of layers, the lesser is the scattering, though this effect is not considerable. The scattering can be considerably reduced by designing an optimum pair design. In case of the optimum pair (OP) design, either one of the H and L, is changed by an amount of about 10% or both are changed by the same amount of ~10%, which results in the displacement of the maximum of the electric field in the multilayer stack from the interfaces to somewhat displaced from them and, therefore, leads to the reduction in absorption and scattering losses. The OP design is given by:

$$\text{Air}[0.9H\ 0.9L]][HL]^{n-1}\text{Glass} \tag{3.2}$$

Further, the electric field is in the multilayer stack from the interfaces, the lesser is the scattering. However, this is accompanied by an increase in the number of layers required for achieving the same reflectivity. Hence, it has to be optimized, and departure of film thicknesses by an amount of ~10%, is good enough.

3.4 Results of Improvements in Scattering Loss by Various Factors

There are certain amounts of variation in the reduction of the scattering loss achieved by various factors, but the approximate values are given below:

1. Clean environment of ~100 class in the cleaning room and the coating room, especially near the coating chamber; and the use of the vacuum pumps not using any oil—~50%
2. The float polishing technique—~40%
3. The optimum pair design—~25%
4. The DIBS technique—~75%.

It is to be noted that each improvement is over the previous figure order wise, and hence the total reduction is a multiple of all these improvement factors. Thus, a scattering loss ~150 ppm (in case of the dielectric laser mirrors) is reduced to ~7–8 ppm, required for the RLG mirrors.

The reduction in the scattering loss of the RLG mirrors (to a level of ~7–8 ppm) has been achieved by using commercially available optical substrates from the markets in USA and Europe, and by using: (i) optimum pair design and (ii) DIBS technology in the rooms of the very high cleanliness levels (100 class). The scattering loss values are on the basis of the scattering measurements made on the total integrating sphere (TIS), and the microroughness measurements made on the Talystep surface profiler, and atomic force microscope. The surface evaluation of the substrates has been done by observing the samples before and after coating, on the Nomarski differential interference contrast microscope.

It is important to note that the improvement in the scattering loss of the laser mirrors is dependent on so many factors, and so it is not possible to have a complete reproducibility in the amount of the achieved improvement. Still, it has been observed on the basis of a number of experiments that it is of the order of ±~10%.

The main advantages of the DIBS technique and a comparison of the coating processes in reducing the reflected wavefront change after vacuum evaporation are illustrated below.

3.5 Advantage of Ion Beam Sputtered Films Over Electron Beam Evaporation Films

- More durable—due to high energy of condensing/sputtered atom. This energy (1–10 eV.) is 10–100 times higher than electron beam evaporation films.

- High mobility of atom
- Higher capacity in filling voids
- A more durable film (Fig. 3.3).

The reduction in the reflected wavefront change after vacuum evaporation for the ion beam sputtering technique (IBST), as measured by the adaptive optics technique [2] is ~5% of the corresponding change for the ion-assisted deposition (IAD) technique, and ~3% of the change for the electron beam gun (EBG) deposition. Therefore, the scattering from the coatings developed by the DIBS unit is reduced proportionately as compared to those developed by the other techniques.

The ring laser gyro (RLG) coatings have been deposited for the mirrors for 30° and 45° AOI, required respectively for the triangular gyro and rectangular gyro, at 0.6328 μm. The values achieved have been: $Rs = Rp \sim 99.95\%$, where *Rs* and *Rp* are, respectively, the reflectivity for the *s*-polarized component of light and the reflectivity for *p*-polarized component of light. Also, (i) the RLG output coupler mirrors with $Ts = 0.1–0.4\%$ or $Tp = 0.1–0.2\%$ (*Ts* and *Tp* being the corresponding transmission values); (ii) the antireflection (AR) coatings with reflection loss/surface as low as 0.02–0.05%; and (iii) the beam splitter coatings with *R*/*T* ratio required for the beam combining prism have been fabricated. A large amount of reduction in the reflection

Reflected Wavefront Change after Vacuum Evap	Coating Process
0.159	E.B.G.
0.102	I.A.D.
0.005	I.B.S.T.

Fig. 3.3 Advantages of the DIBS technique and a comparison of the coating processes in reducing the reflected wavefront change after vacuum evaporation

loss/surface for intracavity elements of the RLG has been achieved (as low as 0.02–0.05%). This has been made possible by: (i) Designing and fabricating our in situ film thickness monitor and fitting it in the Balzers vacuum coating unit, and (ii) making corrections for the moisture penetration in the voids in the structure of the coatings. The choice of the substrate material is very important in case of the optical windows required for the RLG.

3.6 Concluding Remarks

With the increasing interest being shown in various novel applications of the laser systems [3–9], better and more efficient laser mirrors may be required in the future. Therefore, the research and development work on these coatings technologies, may grow in the future, for developing components having coatings with better adhesion, higher resistance to scratch and rubbing, very low absorption (so as to withstand higher temperatures) and scattering, and most importantly higher laser-induced damage threshold (LIDT). JVC-18 is the meeting of the established series of Joint Vacuum Conferences, which are organized at two-year periods by the Austrian, Croatian, Czech, Hungarian, Slovak and Slovenian national vacuum societies. The International Conference on Thin Films is one of the tri-annual conference series endorsed and co-organized by the Thin Film Division of the International Union for Vacuum Science, Technique and Applications (IUVSTA). This merger of the two conferences is an efficient forum for scientists to discuss latest results, exchange ideas and to find partners for collaboration in the field of vacuum science, thin films, and coatings. In this direction, JVC-18 [10] will be held at Budapest, HUNGARY, November 22–26, 2020. It is expected to throw light on various aspects and results of studies on vacuum science and thin films coatings. So this conference is expected to be very useful for the optical engineers and designers engaged in fabricating RLG mirrors and other Intracavity elements for the RLG. Thus, it can be concluded that the field of developing efficient coatings by DIBS technology, is on a firm footing, and is going to evolve.

References

1. K.N. Chopra, A.K. Maini, Thin Films and their Applications in Military and Civil Sectors. Defence Science and Documentation Centre, Metcalfe House, Defence Research and Development Organisation, Ministry of Defence, Delhi, India (2010)
2. K.N. Chopra, A short review on modeling and compensation of the aberrations and turbulence effects by adaptive optics technology. Atti. Fond. G. Ronchi **68**, 579–594 (2013)
3. W. John, High-Power Optics: Laser-Machined Pits Replace Coating Flaws. Laser Focus World, 04/01/2011
4. B.S. Mann, High power diode laser-treated HP-HVOF and twin wire arc-sprayed coatings for fossil fuel power plants. J. Mater. Eng. Perform. **22**, 2191–2200 (2013)

5. Coating High-Power Semiconductor Lasers (Laser Diode Bars) Facets via Ion Beam Deposition. http://www.azom.com/article.aspx?ArticleID=9741
6. Optical Interference Coatings, Optical Technology. Appl. Opt. **50**, 1197–1279 (2011). http://www.opticsinfobase.org/ao/issue.cfm
7. L. Cromer Christopher, E. Hurst Katherine, X. Li, H. Lehman John, Black optical coating for high-power laser measurements from carbon nanotubes and silicate. Opt. Lett. **34**, 193–195 (2009)
8. K.N. Chopra, A technical note on the diode pumped Er fiber lasers and a short review of some important novel studies. Atti. Fond. G. Ronchi **69**, 1–12 (2014)
9. K.N. Chopra, Modeling and designing of the phase conjugated lasers–a technical note. Atti. Fond. G. Ronchi **69**, 691–704 (2014)
10. JVC-18 to be held at Budapest, Hungary, pp. 22–26 (November 2020)

Chapter 4
Improvement in the Laser-Induced Damage Threshold by the Dual Ion Beam Sputtering Technology

4.1 Introduction

The high-power lasers (HPLs) have unique properties—speed of light, no inertia, no "*g*" effects, multishot, quick refiring, fast response rapid retargeting, and high force multiplier, besides being cost competitive which makes them useful for industrial and defense applications. The topic of high-power lasers coatings (HPLCs) is so much intrinsically connected with the HPLs that the progress in this field has followed the advancements made in achieving higher powers in the lasers. It is because of the reason that the lasers for their operation are based on the performance of the mirrors, required for the laser cavity, a useful component in the system. The high-power laser beam has to pass through the output mirror after a number of reflections from the mirrors of the cavity, which have to withstand the power so that the laser operation can continue. As discussed by Chopra and Maini [1], the HPLCs technology is quite complex, and not just based singly on the deposition process. To achieve powers of such magnitude, many pronged attack is required: (i) choice of the coating materials; (ii) choice of the substrate material, (iii) polishing technique for the fabrication of the optical substrate, (iv) deposition technique, and (v) coating design. Besides, the cleaning of the substrate and the overcoats on the mirror are very important for increasing the laser power damage threshold; and also for increasing the adhesion and strength of the HPLCs.

4.2 Parameters of Coatings

It is important to note that some of the points discussed here are the same as presented in the Chap. 3 but are given for the sake of continuity. The HPL coatings are fabricated by considering various parameters—(i) coating material, which should be pure and must have absorption edges away from the wavelengths of interference and also of the harmonics (for pulse irradiation); (ii) coating technique should be able to give

K. N. Chopra, *Optoelectronic Gyroscopes*, Progress in Optical Science and Photonics 11,
https://doi.org/10.1007/978-981-15-8380-3_4

high packing density to the coatings, so that there is no scope for the voids in the films in which water and chemicals may diffuse resulting in changing the optical thickness of the coatings, and also reducing the mechanical strength of the coatings, and therefore their damage threshold; (iii) coating design, which should produce shifting of electric field maximum away from boundaries of the multilayer stack, i.e., interfaces between the different layers, where the impurities tend to concentrate. The achievement of the high-power damage threshold for the multilayer dielectric coatings for laser systems depends on thin films design (having minimum number of layers), so that the absorption and scattering from the layers and the interfaces are minimized. For this, the materials for the high- and low-refractive indices should be so chosen that their difference is maximized, which results in minimizing the number of layers for achieving the same results of reflection. The choice of the substrate material is especially important in case of the optical windows required in the laser systems. The choice of the coating material is also very important, from the point of view of considering various parameters, e.g., thermal conductivity, linear expansion coefficient, and good Poisson's ratio, besides having high degree of scratch resistance and excellent polishing property; and (iv) polishing technique used for polishing the substrate—the usual polishing techniques of abrasives produce a smooth surface of quite good quality, but not sufficient for developing the HPLCs. Any abrasive material left in the substrate at some scratch or dig becomes a center of absorption, where the coating may go bad due to excessive heating. Also, the size of the particles cannot be below some microns level, and so the surface has some root mean square value of roughness 10–15 A°, and not lower, which becomes the source of scattering and absorption. Therefore, to overcome this problem, float polishing technique (for optical substrates) and single point diamond turning machine (for metallic substrates) are used, and they are able to give the surface rms values ~2–3 A°, and hence lead to considerable improvement in the laser damage threshold of the coatings. These techniques are good for the metal surfaces and IR surfaces like ZnSe or KCl, since both these techniques give very smooth surfaces, nearly free from the trapped material, which are likely to increase the absorption by an amount proportional to the refractive index of the trapped material.

The most important factor for increasing the LIDT is that of the choice of the coating technique employed for the fabrication of the coatings. Out of all the coating techniques available like—electron beam (EB) evaporation, sputtering, ion-assisted deposition (IAD), and the dual ion beam sputtering (DIBS) technique, the last one which is a mixture of ion beam sputtering and ion-assisted deposition is the best. In DIBS technique, one ion beam is used for sputtering the coating material from the target inside the vacuum chamber, and the other is used for assisting the deposition on the substrate. Due to the high energy of the adatoms (300 eV), as compared to ~1 eV in case of E.B, the adatoms move on the surface till they find unfilled space, and thus do not form the columnar structure, but give rise to a smooth surface, having packing density ~1, unlike ~0.9 in case of the EB evaporation, and 0.93 in case of sputtering techniques, which as a consequence do not leave voids for moisture or chemicals from atmosphere to diffuse through, and hence protect them and help them in achieving very high damage threshold, in addition to a very high degree of adhesion. The optical substrate should have low absorption, e.g., Si, W, and Cu. Also, the material should be free from color centers, subsurface damage, inclusions, and nodules, because each of these reduces the damage threshold. As mentioned earlier, the cleaning of the substrate and the deposition of the overcoats on the mirror also play important roles for increasing the laser power damage threshold; and also increasing the adhesion and strength of the HPLCs. The cleaning is more of an art than science and contributes very largely for achieving high-power laser damage threshold, and adhesion/strength of the coatings. The process is controlled critically by many parameters including the rate of deposition, substrate temperature, oxygen partial pressure in case of the designs including dielectric metal oxides, thickness calibration, and the material-melt preconditioning. The optimization of these process control parameters determines the stoichiometry of the growing film and helps in producing a homogeneous layer with the desired metal-oxygen content and structure. The materials like TiO_2 have a tendency to break into lower oxides—TiO, and so the flow rate of O_2 has to be optimized while the coating of such materials. In addition to all these parameters, deposition of all the HPLCs, are done by giving planetary rotation (~30 rpm) to the substrate, for giving uniform LIDT, and also annealing of the coatings (at 300 °C), which has also been observed to increase the LIDT considerably (~15%).

Designing of a multilayer stack is based on the quarter wave stack consisting of the layer sequences as given below:

$$[\mathrm{Air}][HL]^n\mathrm{Glass} \tag{4.1}$$

where H denotes a high index layer of quarter wave optical thickness (QWOT), L denotes a low index layer of quarter wave optical thickness (QWOT), and $[HL]$ denotes the basic period, which is repeated n times. The reflectivity increases with increasing n and becomes very high indeed (~99.9%) for $n \sim 15$. It should be noted that the value of n required for a particular reflectance is minimized by maximizing the difference in H and L.

For optimum pair (OP) design, either one of the H and L, is changed by about 10% or both are changed by an amount of ~10%, which results in the displacement of the maximum of the electric field in the multilayer stack from the interfaces to somewhat displaced from them, and therefore leads to the reduction in absorption and scattering losses. The OP design is given by:

$$\mathrm{Air}[0.9H\ 0.9L]][HL]^{n-1}\mathrm{Glass} \tag{4.2}$$

4.3 Coatings for Some Types of HPLs

Some of the common types of HPLs are: gas dynamic laser (GDL), hydrogen fluoride laser (HFL), and chemical oxygen iodine laser (COIL). The HFL optics and thin film coatings for (2.6–3.5 μm—fundamental) and (1.33 μm—overtone) are:

1. Mirrors: Be–Cu, silicon, Ni-coated copper Cr + Ag + (ThF_4/ZnSe) (base), for protection and enhancement (R ~ 99.2%); high damage threshold ~120 kw/cm^2.
2. Window/couplers: KCL, CaF_2, BK-7, fused silica, and ZnSe.

The coating materials used are: ThF_4, TiO_2, and SiO_2 for high damage threshold ~120 kw/cm^2. The polishing techniques used for these mirrors are: single point diamond turning (SPDT), pitch polishing, cloth + diamond paste (or linda). The required optical finish is: flatness—$\lambda/10$, scratch/dig 30/20, roughness—15–20 Å for fundamental mode; and $\lambda/10$, scratch/dig 30/20, roughness—15–20 Å for overtone.

4.3.1 *Thin Film Coatings for Gas Dynamic Laser (GDL) Optics (10.6 μm)*

Thin film coatings for GDL optics at 10.6 μm are:

Mirrors (flats and curved): Silicon, OFHC copper electroless nickel on copper, Be–Cu, and Moly. Cr + Ag/Au. The reflectivity obtained without coating is: ~99.5%, ~99.4%, and ~98.2%, respectively, with damage threshold >120 kw/cm^{2}cont, 130 kw/cm^2, and 200 kw/cm^2. ThF_4 required for HF mirrors is radioactive, and MIRA material is its suitable replacement, processed, and made available by M/s Balzers, Liechtenstein, Switzerland. The polishing requirements are: optical flatness ~$\lambda/40$, scratch/dig—40/20, and the microroughness—30 Å.

4.3.1.1 Figure of Merit (FOM)

Figure of merit is an important consideration for the choice of the substrate material and is given by:

$$F = \frac{K}{B_e \alpha (1 + \nu)} \tag{4.3}$$

where K = thermal conductivity, B_e = effective absorption, α = linear expansion coefficient, and ν = Poisson's ratio.

The mechanical constants for mirror substrates—Si, Cu, and Mo are very important. The respective values (for these metals) of K, α, and ν are 1.56, 3.90, and 1.33 (W/cm/°C); 2.56, 16.6, and 5.4 (10–6/°C); and 0.28, 0.34, and 0.28 (unit less). On the basis of these values, Si is ~2 times better than Cu or Mo, which have nearly same FOM. These substrates can withstand power densities ~1 kW/cm^2.

It is very important to have this characteristic, as otherwise, it results in: (i) distortion of the optical surface, (ii) loss of mode control, (iii) loss of output power, (iv) thermal focusing in the output beam, and (v) sometimes the damage to whole system.

It is also very important to reduce absorption, which can be done by: (i) Choosing coating materials for the multilayer stack, with higher difference in their refractive indices, as it leads to the requirement of the lesser number of layers and so lesser absorption; (ii) designing a low E-field ML-stack, which is done by adjusting the layers thicknesses so that maximum value of the electric field, Emax occurs in the material with lowest absorption rate. The optimum pair design is able to increase the laser-induced threshold damage (LITD), by an amount ~30%, the amount of increase depending on the difference of the layer thicknesses from the quarter wave thicknesses. In fact, farther the thickness of the layer from the quarter wave thickness, the larger is the increase in the LITD. However, it has to be noted that it also leads to an increase in the required number of layers, and hence more absorption and scattering and, consequently, reducing the LITD. Thus, the optimum pair design has also to be optimized for the best possible increase in the LITD, which has been observed to be ~ 30%. Besides, it is important to control the other damage mechanisms—absorption by defects in thin films and absorption by impurities included in thin films.

4.3.2 Improved Film Stability by Ion Beam Sputtering Deposition

The conventional evaporation provides films with columnar microstructure (which has been confirmed by the electron microscopy studies) and with voids and low packing density. These are also controlled to some extent by the deposition and process control parameters like the rates of evaporation, degree of vacuum, flow rate of the reactive gases (O_2 in case of oxides), and the temperature inside the coating unit, etc. These voids are the centers through which the moisture penetrates over a long duration of time, which leads to changes of the film index of refraction, and therefore the optical thickness.

4.3.3 Effect of Overcoat Layer on High Reflection Coatings

It has been verified that a 15-layer coating of TiO_2/SiO_2 stack with a SiO_2 overcoat results in increasing the LITD by ~40%. This is explained by: (i) lower net residual stress, (ii) greater mechanical strength, and (iii) prevention of adverse chemical reaction with the atmosphere. The overcoat is of half wave thickness, and hence does not change the reflectance and transmittance of the coatings; and thus behaves as an absentee layer for the spectral characteristics.

4.3.4 Advantages of IBSD Over Plasma Sputtering (RF or Magnetron) and Electron Beam Deposition

The IBSD technique is advantageous in the sense that there is no direct exposure of substrate to plasma, and so the intense bombardment by electrons and ions from the plasma is avoided. Also, the deposition is carried at a low pressure (<10–2 Pa), which results in minimizing the contamination of the growing films and the gas incorporation. The other advantage is that the deposition parameters, e.g., ion energy and the angle of incidence at the target are precisely controllable, and also independent of each other. This technique gives the stable optical properties due to the non-columnar structure. Also, the process results in: (i) lower extinction coefficient due to the lower absorption losses; (ii) stability in the adverse environments, e.g., high temperature and presence of reactive gases, etc.; (iii) higher P.D. and so higher refractive index and hence higher stability with time; (iv) lower grain size and so lesser scattering; (v) better adhesion, because of comparatively much larger energy of the adatoms; (vi) reduction in stress; and (vii) better stoichiometry. Some of the advantages of the DIBS technique and a comparison of the coating processes in reducing the reflected wavefront change after vacuum evaporation are illustrated below.

4.3.4.1 Advantage of Ion Beam Sputtered Films Over Electron Beam Evaporation Films

Advantage of ion beam sputtered films over electron beam evaporation films have been shown in Fig. 4.1.

- More durable—due to high energy of condensing/sputtered atom. This energy (1–10 eV.) is 10–100 times higher than electron beam evaporation films.

- High mobility of atom
- Higher capacity in filling voids
- A more durable film.

Reflected Wavefront Change after Vacuum Evap	Coating Process
0.159	E.B.G.
0.102	I.A.D.
0.005	I.B.S.T.

Fig. 4.1 Some of the advantages of the DIBS technique, and a comparison of the coating processes in reducing the reflected wavefront change after vacuum evaporation

It can be observed that the ion beam sputtering technique is able to produce very stable coatings in which the reflected wavefront change after vacuum evaporation is reduced to ~5% or even <5% as compared to the other processes.

4.3.5 DIBS Technology for the High-Power Laser (HPL) Coatings

The DIBS technology is based on utilizing two ion beam guns, one for sputtering the target, and the other for assisting the deposition. The working and the schematic of the process are explained by the following figure (Fig. 4.2).

The output coupler with the required transmission ($T \sim x\%$) is fabricated by: (i) Designing and fabricating in situ film thickness monitor capable of measuring reflection/transmission during coating.

However, it has to be noted that before the deposition of the coatings for the HPLs, the substrates have to be cleaned thoroughly, and inspected under differential interference contrast (DIC) microscope, atomic force microscope (AFM), and also by the Fourier transform infrared (FTIR) technique. In fact, these tools are also

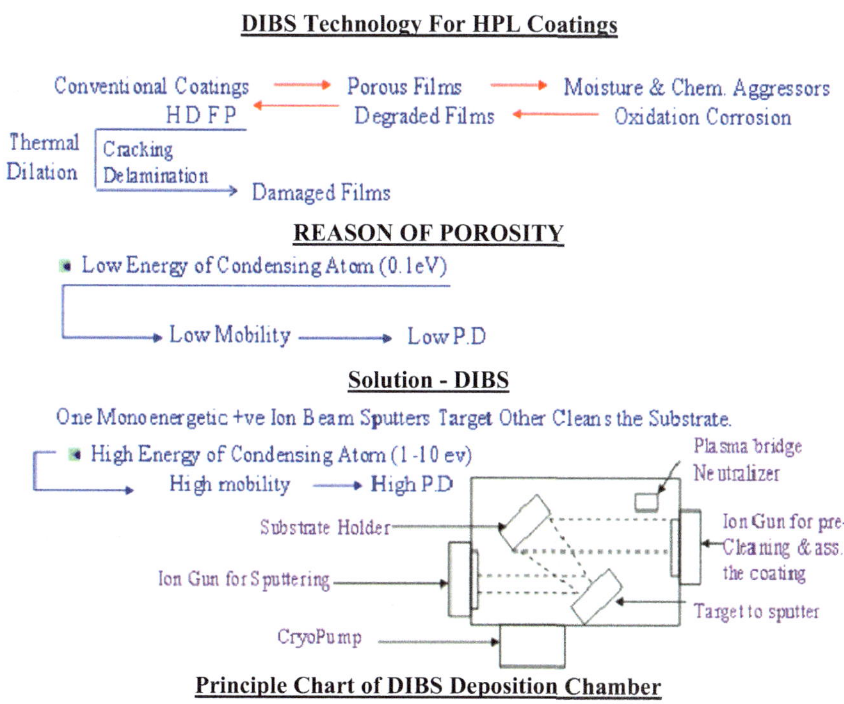

Fig. 4.2 Schematic of DIBS technology and the working of the process

used for the optimization of the process control parameters, by testing the coated samples after each variation of the parameters. Differential interference contrast microscopy, also known as Nomarski interference contrast (NIC) microscopy or simply as Nomarski microscopy, is indeed an important optical microscopy illumination technique used for enhancing the contrast in the unstained and transparent samples. Atomic force microscopy, also known as scanning force microscopy (SFM) is a very high-resolution type of scanning probe microscopy, with resolution ~ fractions of a nanometer, i.e., more than 1000 times better than the optical diffraction limit. Fourier transform infrared is a technique which is used to obtain an infrared spectrum of absorption, emission, photoconductivity or Raman scattering of a thin film, and is really useful in detecting the presence of the unwanted groups (Si–O, OH, etc.) in the films.

The clean room facilities and the types of pumps are also very important parameters for the successful fabrication of these coatings. As discussed in the last chapter, clean room conditions are necessary because these reduce the risk of recontamination after cleaning the substrates, and while loading them in the coating chamber. It is a common practice to use the lint-free wipes without silicone constituent for cleaning manually in the final cleaning process. Mostly, the solvents used are—methanol, isopropanol, or acetone, and the ultrasonic cleaning is found to be useful and more effective at dislodging the residual polishing compounds than cleaning by hand. Of course, now, the float polishing technique is used for such coatings, and so the role of this cleaning is reduced. Also, the drag-wipe technique is used to produce very high shear forces, sufficient for the removal of any remaining contaminants from the surface. Utmost care has to be taken to stop the migration of the contaminants from several parts of a coating chamber onto the optical surfaces. Backstreaming can occur from the diffusion pump, resulting in organic contamination, and hence for HPLCs, their use is totally avoided. Instead, the pumps like—turbomolecular pumps, based on the principle that gas molecules can be given momentum in a desired direction by repeated collision with a moving solid surface, which is done a rapidly spinning turbine rotor hitting the gas molecules from the inlet of the pump toward the exhaust to create or maintain a vacuum, and cryopumps, also called cryogenic pumps, which are the vacuum pumps that trap gases and vapors by condensing them on a cold surface, are employed. The walls of the chamber can also contribute to the contamination of the substrates, as the material evaporated from the target deposits on the substrates, and also on the walls of the chamber, which after many runs, has a tendency to flake off. To get rid of this problem completely, it is customary to put a lining of the aluminum foil or to have coaters using removable steel liners. It is important to note that the improvement in the LIDT of the laser mirrors is dependent on so many factors, and so it is not possible to have a complete reproducibility in the amount of the achieved improvement. Still, it has been observed on the basis of a number of trails that it is of the order of $\pm\sim 10\%$.

4.4 Concluding Remarks

A large number of studies on the HPLs and the HPLCs have recently been made, and some of the recent novel studies are discussed here. Mann [2] has studied the high-power diode laser (HPDL) surface modification of twin wire arc-sprayed (TWAS) and high-pressure high-velocity oxy-fuel (HP-HVOF) coatings to combat solid particle erosion occurring in fossil fuel power plants. They have overcome the solid particle impact wear above 673 K, by using the Cr_3C_2-NiCr-, Cr_3C_2-CoNiCrAlY-, and WC-CrC-Ni-based HVOF coatings. By using robotically controlled parameters, they have carried out the HPDL surface treatments of TWAS-coated high strength X10CrNiMoV1222 stainless steel and HP-HVOF-coated AISI 410 stainless steel samples and have compared these with the HPDL-treated high chromium stainless steels and titanium alloy for high energy particle impact wear (HEPIW) resistance. It has been reported that the HPDL surface treatment of the coatings has improved the HEPIW resistance manifold. It has been discussed and reported [3] that the laser diode bars come under the category of high-power semiconductor lasers and are constructed by joining several single emitters having a width of 100–200 μm in an array, in which each diode bar generally consists of 20 and 50 emitters. It has been emphasized that a typical commercial device can generate tens or even several hundreds of watts of output power, and therefore, these high-power diode arrays are used in pumping of solid-state laser systems for medical, military, commercial, and industrial applications in addition to material processing applications such as surface treatment, cutting, and welding. Wallace [4] has made a very interesting study and has reported that intentionally fabricated conical flat-bottomed pits have damage thresholds much higher than the coating flaws they replace, and thus allow the large optics with imperfect coatings to be used in high-peak-power environments. A special issue [5] on the optical interference coatings has appeared, which has many useful papers on the optical interference coatings and their characterization. More interest [6–8] has been shown recently in the HPLs and HPLCs. Cromer et al. [6] have described a coating based on potassium silicate, commonly known as water glass, and multiwall carbon nanotubes, which has a high absorbance (0.96 at 1064 nm in wavelength) and a laser damage threshold that is comparable to that of ceramic coatings presently used for commercial thermopiles for high-power laser measurements. They have documented the coating process and experiments that demonstrate that the laser damage threshold at 1064 nm is 15 kW/cm^2. PolarOnyx laser's proprietary technology on fiber lasers [7] has opened a variety of applications in many fast-growing industries like material processing, bio-medicals, sensors, and optical communications, because of the unique competitive advantages of the technology in the sense of providing the capability of developing high energy/power ultrafast pulsed, compact and robust fiber laser products, which are maintenance-free and cost-effective for both industrial and scientific applications. Chopra [8] has given the technical analysis of the diode-pumped Er-fiber lasers and has also reviewed their important novel studies. Epperlein et al. [9] have written a reference book, which provides a fully integrated novel approach to the development of high-power, single-transverse mode, edge-emitting diode lasers by addressing the complementary topics of device engineering, reliability engineering, and device diagnostics. They have also discussed the diode laser fundamentals, followed by an elaborate discussion of problem-oriented design guidelines and techniques, and by a systematic treatment of the origins of laser degradation and a thorough exploration of the engineering means to enhance the optical strength of the laser and have reviewed the novel, advanced diagnostic methods. With the progress achieved in the deposition techniques, laser engineers are concentrating their efforts on using the new advanced materials like quantum dots and nanoparticles, which may allow greater laser control. However, the coatings are expected to continue to play an important role in the laser applications.

It can be concluded that the IBSD technique is advantageous in the sense that it (i) gives the stable optical properties due to the non-columnar structure; (ii) results in the lower extinction coefficient due to the lower absorption losses; (iii) provides stability in the adverse environments; (iv) gives higher P.D. and so higher refractive index and hence higher stability with time; (v) results in lower grain size and so lesser scattering; and (vi) gives better adhesion and better stoichiometry. Also, the technique is able to produce very stable coatings in which the reflected wavefront change after vacuum evaporation is reduced to ~5% or even <5% as compared to the other processes. In addition, a 15-layer coating of TiO_2/SiO_2 stack with a SiO_2 overcoat results in increasing the LITD by ~40%. The optimum pair design is able to increase the LITD, by an amount up to ~30%, depending on the difference of the layer thickness from the quarter wave thickness. It can be safely stated that the performance of this technique, for increasing the LIDT, is such that it is able to achieve total improvement of ~120%, by taking all the steps discussed above. It has also been observed in our investigations that Ta_2O_5 (a good alternative to TiO_2) is another coating material which is very good for high-power laser coatings. In fact, it has been observed that there is an increase in the laser-induced damage threshold of ~10–15% over the value achieved by using TiO_2. Also, in view of the novel studies on the high-power lasers being undertaken, the subject of high-power coatings is expected to grow and evolve.

References

1. K.N Chopra, A.K Maini, Thin Films and their Applications in Military and Civil Sectors. Defence Science and Documentation Centre, Metcalfe House, Defence Research and Development Organisation, Ministry of Defence, Delhi, India (2010)
2. B.S Mann, High power diode laser-treated HP-HVOF and twin wire arc-sprayed coatings for fossil fuel power plants. J. Mater. Eng. Perform. **22**, 2191–2200 (2013)
3. Coating High-Power Semiconductor Lasers (Laser Diode Bars) Facets Via Ion Beam Deposition. http://www.azom.com/article.aspx?ArticleID=9741
4. J. Wallace High-Power Optics: Laser-Machined Pits Replace Coating Flaws. Laser Focus World (04/01/2011)
5. Optical Interference Coatings. Opt. Technol. Appl. Opt. **50**, 1197–1279 (2011). http://www.opticsinfobase.org/ao/issue.cfm

6. C.L. Cromer, K.E. Hurst, X. Li, J.H. Lehman, Black optical coating for high-power laser measurements from carbon nanotubes and silicate. Opt. Lett. **34**, 193–195 (2009)
7. Onyx Polar Laser's proprietary technology on fiber lasers. PolarOnyx Laser (2014). http://www.polaronyxlaser.com
8. K.N. Chopra, A technical note on the diode pumped er fiber lasers and a short review of some important novel studies. Atti. Fond. G. Ronchi **69**, 299–310 (2014)
9. P.W. Epperlein, *Semiconductor Laser Engineering, Reliability and Diagnostics: A Practical Approach to High Power and Single Mode Devices Hardcover* (Wiley Ltd., UK, 2013)

Chapter 5
Optical Testing of Optical Elements

5.1 Introduction

Optical Testing of Optical Elements is of utmost importance in the optical instrumentation. The pioneering work on providing the simplest method for testing an optical instrument was done by the Italian physicist Ronchi [1, 2], who in 1923 invented the technique, named as "Ronchi test." The Ronchi tester consists of a light source, a diffuser, and a Ronchi grating, composed of alternate dark and clear stripes. One design is a small frame with several evenly spaced fine wires attached. The basic configuration of the Ronchi test and the shapes of the related fringes are shown in Fig. 5.1.

As is clear, the light is emitted through the Ronchi grating, or a single slit, and after reflection by the mirror being tested, passes through the Ronchi grating (placed at a short distance <2 cm closer to the mirror) again, and is then seen by the observer's eye, placed close to the center of curvature of the mirror under test, looking at the mirror through the grating. The observer sees the mirror covered in a pattern of stripes that reveal the shape of the mirror and analyzes the pattern by comparing it with a mathematically generated diagram (usually done on a computer today) of what it should look like for a given Fig. 5.2.

The observer sees the mirror covered in a pattern of stripes that reveal the shape of the mirror. Obviously, the inputs to the program are line frequency of the Ronchi grating, focal length, the diameter of the mirror, and the required figure. The pattern consists of straight lines in the case of a spherical mirror. In optical testing, the Ronchi test is a method of determining the surface shape of the mirrors [3, 4], especially concave mirrors, used in telescopes and other optical devices. It provides a quick observation of the mirror's shape and condition and also the identification of the so called "turned edge," i.e., the rolled down outer diameter of the mirror, a common fault developed in the objective mirror making. The interferometer based on the Ronchi test is named as "Ronchi interferometer," which is simply composed of an optical bench and a low-frequency optical grating—Ronchi grating.

K. N. Chopra, *Optoelectronic Gyroscopes*, Progress in Optical Science and Photonics 11,
https://doi.org/10.1007/978-981-15-8380-3_5

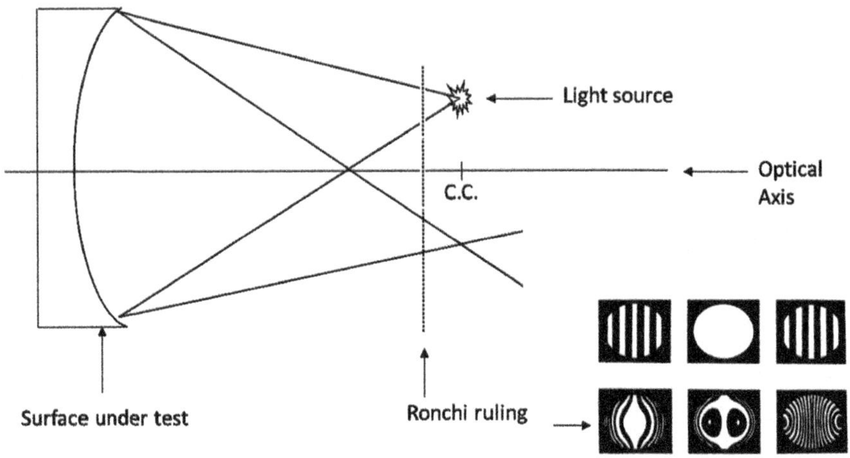

Fig. 5.1 Basic configuration of Ronchi test and the shapes of the related Fringes. *Figure courtesy* opticalengineering.spiedigitallibrary.org

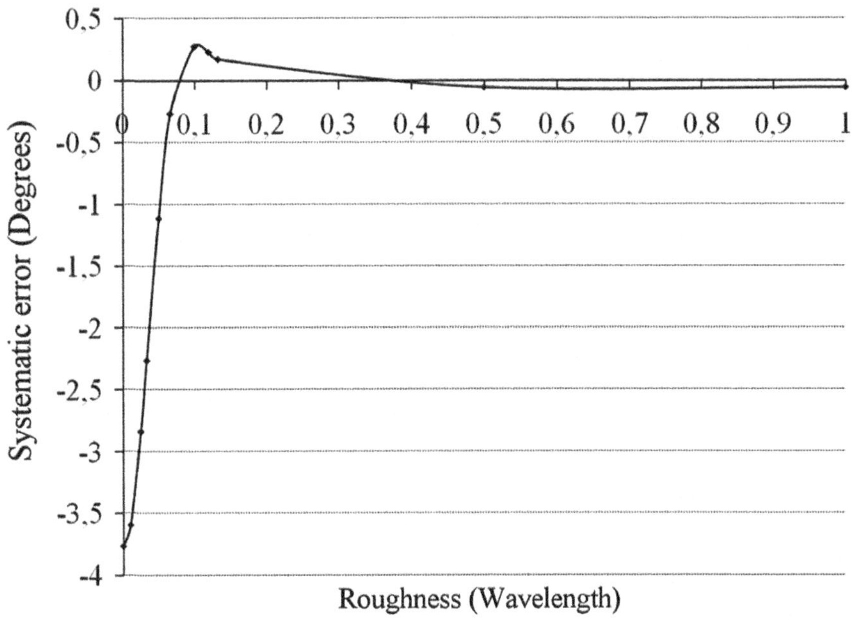

Fig. 5.2 Variation of systematic error with roughness (wavelength). *Figure courtesy* opticalengineering.digitalliobrary.org

Recently, a new phase reduction algorithm [5], originally designed for interferometry test of large optics, has been presented for the Ronchi test. As discussed by Wan and Lin [5], a good automatic phase reduction algorithm is necessary to fully interpret the Ronchigram. It has been shown that due to the common path property, only two Ronchigrams shifted by $\pi/2$ are necessary for reproducing the phase. Accuracy better than one-thirtieth of the grating space has been reported. In addition, the methods have been suggested for finding the spherical aberration and astigmatism without integrating the phase using the Ronchi test. Interestingly, the Tests of results using the new algorithm have shown good agreement with typical interferometry tests, thereby illustrating the importance of Ronchi test even today, in the modern advanced methods for finding the spherical aberration and astigmatism, very commonly encountered in the optical systems.

There are many types of optical testing: optical non-destructive testing (NDT), improved interferometric optical testing, optical time-domain reflectometer, optical interferometry, and fiber optic measurement technique. These techniques are of tremendous utility in making any important studies related to various branches of optics and optoelectronics including imaging, interferometry, and laser systems. Each is characteristic for a particular application and is therefore selected accordingly. A technical analysis of some of the important techniques has been done and discussed below.

5.1.1 Optical Non-destructive Testing (NDT)

Optical non-destructive testing (NDT) has drawn the attention of many workers in recent years [6–8], because of its non-destructive imaging characteristics with high precision and sensitivity.

5.1.1.1 Laser Speckle Inteferometry and Speckle NDT

Muravsky et al. [9] have presented a two-step electronic speckle pattern interferometry (ESPI) method with blind phase shift of a reference wave, in which the phase-shift extraction is fulfilled by using both—the correlation and the discrete Fourier transform (DFT) approaches. In the correlation approach, the blind phase shift is calculated via the correlation coefficient between two similar speckle interferograms (SI) differing only by the reference wave phase shift. The DFT approach is based on the blind phase-shift extraction from frequency components of SI Fourier spectra. Muravsky et al. [9] have testified to the high performance of both these approaches by analyzing and comparing them. Moreover, it has been shown that the correlation approach is preferable for blind phase-shift extraction from SIs with a rough surface than from interferograms with a smooth surface, thus making it more convenient for ESPI than for phase-shifting interferometry. It has been emphasized that this approach provides a lesser level of systematic error of extracted phase shift in comparison with the DFT. The results for the dependence of systematic error on roughness (wavelength) are available in the literature and have been reproduced below.

5.1.1.2 Infrared Thermography NDT

The technique of the infrared thermography NDT is based on IR thermography (a technique for producing an image by using the infrared light emitted by objects due to their thermal condition) and has many advantages: (1) being fast, and (2) providing non-contact, non-interaction, and real-time measurements over a large detection area with a long range, which result in its use for applications in many fields including aviation, aerospace, machinery, medical, petrochemical, and power. The technique uses thermography camera, a device which produces a live video picture by using infrared radiation intensities from a surface, and then converting the spatial variations in infrared radiance into a two-dimensional image, displaying the variations in radiance as a range of colors or tones. It is important to note that infrared thermography is based on the temperature difference, the objects in the image lighter in color are warmer, and darker objects are cooler, and that the image produced is called a thermograph.

Thermography is of two types: (1) active thermography (AT) defined as the application of a stimulus to a target to cause it to heat or cool in such a way, which allows the observation of the characteristics (flaws sought in NDT or norms sought in quality control) of the target when viewed by thermal imagery and (2) passive thermography (PT) defined as measurement of the temperature differences between the target materials and the surroundings under different ambient temperature conditions. Active thermography, on the basis of different heating methods, can be further divided into pulsed thermography (PT), pulse-phase-thermography (PPT), modulated thermography (MT), and ultrasonic lock-in thermography (ULT). A lot of progress has recently been made in the field in the form of (1) the integration of infrared and other testing methods, (2) the change of signal processing method from amplitude information to phase information, and (3) the increased stress on field testing oriented systems. An experimental setup for the lock-in thermography in reflection with optical excitation is given in Fig. 5.3.

The thermal waves affected by the internal defects are picked up by the thermal camera and are studied by the control unit and the PC.

5.1.1.3 Terahertz (THz) NDT Technology

The NDT is useful for detecting the barely visible damage with optical NDT methods like—digital shearography, ESPI, and infrared thermography—which compared to the traditional NDT methods provide many advantages: resistance to electromagnetic interference, non-contact, and the real-time results, which have been established in many applications in the laboratory, factory, and field environment, because of

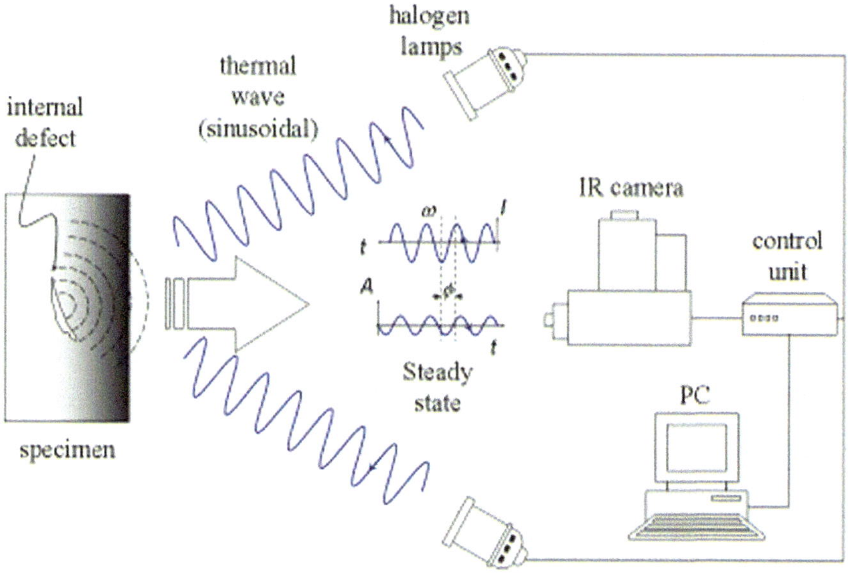

Fig. 5.3 An experimental setup for the lock-in thermography in reflection with optical excitation. *Figure courtesy* www.visiooimage.org

providing higher sensitivity and resolution for high-precision applications. Presently, optical NDT techniques are widely used in areas like aerospace, civil, medical, and petrochemical, which explains the emphasis on the further optical NDT technology research including the development of integrated electromagnetic NDT imaging systems. The experimental setup for the pulsed terahertz TDS system is shown in Fig. 5.4.

Pulsed terahertz waves are provided by the ultrafast laser, which on being split by the optical beam splitter, traverse different paths—one through the scanning optical delay line and the other through the multilayered structure with defect—and then combine subsequently to give an output, which is analyzed for the study of defects.

5.1.2 Interferometric Optical Testing

Interferometry is widely used as a powerful tool for testing optical components and optical systems and also as the metrology of many other components like flatness and roughness of hard disk drive platters and the shape of magnetic recording heads and machined parts. Optical interferometry is an optical measurement technique, which provides extreme precise measurements of distance, displacement or shape and surface of objects, by exploiting the phenomenon of light waves interference, where under certain conditions a pattern of dark and light bars called interference

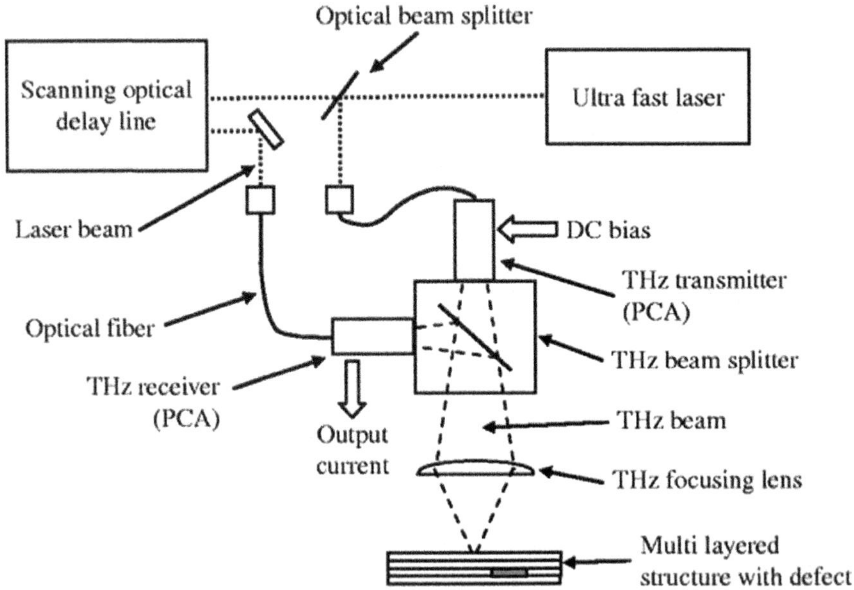

Fig. 5.4 Scheme of pulsed terahertz TDS system. *Figure courtesy* www.emeraldinsight.com

fringes can be produced, that can be analyzed to present accurate measurements in the range of nanometer. Optical interfrometry is an optical instrument that can produce two beams interference or multiple beam interference by (1) wavefront division, in which two light beams from the same wavefront are made to interfere to produce an interference fringe pattern and (2) amplitude-division, in which a light beam from one source point is divided into two beams using a beam splitter. With the advent of the recent developments in laser, fiber optics and digital processing techniques have supported optical interferometry, leading to the newer applications ranging from the measurement of a molecule size to the diameters of stars. Some important advances have resulted in increasing the area of the applications of interferometric metrology, which are: (1) a technique for reducing the sensitivity of interferometry to vibration; (2) a method for measuring deformations and vibrations of diffuse surfaces; and (3) a method for reducing the sensitivity to vibration and coherent noise and hence enabling the measurement of parallel glass surfaces. These advances have resulted on the basis of the phase-shifting interferometry, which has enhanced the utility and effectiveness of interferometry in some important industrial and research applications. Some important research studies [10–13] on various topics like pixelated phase-mask dynamic interferometer, the geometric phase, dynamic phase-shifting electronic speckle pattern interferometer, and basics of interferometry have been reported. Some of these techniques are technically discussed below.

5.1.2.1 Phase-Shifting Interferometry

Phase-shifting interferometry is the tool for feeding the interferometric data into computers, based on three unknown parameters—the amplitude of the reference beam, the amplitude of the test beam, and the phase difference between the two interfering beams, the last one being of paramount interest, since it gives the optical path difference, which helps in determining: (1) measures to be taken for the correction of optics and (2) the performance of optics in the absence of the application of this correction. The phase difference between the two interfering beams, which is changed in a known manner—by 90° between consecutive intensity measurements, is determined by measuring the intensity of the interference fringes. At least three intensity measurements have to be made to find the values of the three unknown parameters. It is convenient to use ninety degree phase steps in order to simplify the calculations, because the sines and cosines of the phase difference are being measured. The interference fringes are detected by a solid-state detector array, the output of which is digitized, the digitized data being fed directly into computer memory. The phase difference between the two interfering beams is changed by a phase shifter, e.g., moving reference mirror or an electro-optic modulator. The computer controls the phase difference between the two interfering beams by reading the detector output. By using various algorithms, the phase difference is computed from these fast and accurate measurements, so that vibration or air turbulence may not change the phase difference between the two beams in unknown ways, thereby introducing large errors.

5.1.2.2 Phase-Shifting Single-Shot Interferometer Technique

The phase-shifting single-shot technique is useful for reducing the sensitivity to vibration and also for the complete data acquisition in a single laser pulse. Though there are many types of phase-shifting single-shot interferometers, the most useful and commonly employed approach, which works well over a large spectral bandwidth, is based on using a quarter-wave plate followed by linear polarizer at different angles to measure the phase shifts. Interestingly, the test and reference beams have opposite circular polarizations, which on transmission through a linear polarizer result in a phase shift between the two interfering beams proportional to twice the rotation angle of the polarizer. A phase shifter of this type is often called a geometrical phase shifter, since the phase shift is independent of the wavelength. It is important to note that the micropolarizer phase-shifting array interferometer is able to work well in the presence of vibration and with a wide range of source wavelengths, because of the fact that the interferometer makes short exposures, and consequently, the vibration and the air turbulence are nearly frozen. Another approach of reducing the effects of air turbulence is that of taking many sets of data, arranging for the time between taking the datasets to be larger than the time taken for the turbulence to change, and then taking the average of the data.

In single-fringe measurements, the theory of quantum metrology has focused on the properties of the input state, and hence, it is often implicitly assumed that the optimal measurement strategy for a given input state can be implemented with available technologies. Therefore, the technical challenges involved in the experimental realization of the output measurement have still been able to receive lesser attention than that in case of the problem of quantum state preparation. However, it is considered that an experimental demonstration of phase sensitivity is not complete without the proper output measurement. It is now well established that for the NOON states of any photon number (NOON states are of particular interest in quantum metrology), the optimal phase sensitivity can be obtained from a two-outcome measurement which assigns a value of $+1$ to even photon numbers in the outputs and -1 to odd photon numbers in the output, and ideally, this parity measurement can be performed by a quantum circuit which identifies the identity of the number of photons at one output being odd or even, without the need of actually counting them, which, at present, is not possible, because in the absence of unit-efficiency photon-number-resolving detectors, this is really very difficult to realize. However, for the case of the efficient posts, election of a particular photon number can be more easily realized with high, but non-unit, efficiency detectors. In the existing technique, demonstrations with $N > 2$ have relied on projecting the output of the interferometer onto just one or two states of definite photon number in each output arm, e.g., $|31\rangle$ or $|13\rangle$ in the case of a four-photon experiment. In fact, the results of a single projection of this kind are referred to as a "single fringe," which implies the phase-dependent projection probability for one pattern of photon counting at the interferometer's outputs. As the single-fringe measurements test only a single "yes/no" condition, the experimental requirements for their realization are much simpler than those for the realization of a detection stage, which can separate all $(N + 1)$ photon number distributions in the output.

In addition, the advantage of single-fringe measurements is that the phase sensitivity is directly related to the resolution of the characteristic features in the experimentally determined fringe. It has been well understood that the efficiency of the measurement must be taken into account, when trying to demonstrate phase sensitivities beyond the shot noise limit (SNL), and single-fringe analysis can help to establish a very general relation between intuitively accessible features of multiphoton interference and fundamental issues of phase sensitivity. Quantum entanglement can help to increase the precision of optical phase measurements beyond the SNL to the ultimate Heisenberg limit. However, the N-photon parity measurements required to achieve this optimal sensitivity are extremely difficult to realize with the current photon detection technologies, requiring high-fidelity resolution of $(N + 1)$ different photon distributions between the output ports. Recent experimental demonstrations of precision beyond the SNL have, therefore, used only one or two photon-number detection patterns instead of parity measurements. Xiang et al. [14] have investigated the achievable phase sensitivity of the simple and efficient single interference fringe detection technique and have shown that the maximally entangled "NOON" state does not achieve optimal phase sensitivity when $N > 4$, rather than that the Holland–Burnett state is optimal. They have experimentally demonstrated this enhanced sensitivity using a single photon-counted fringe of the six-photon Holland–Burnett state. It has been claimed that the single-fringe six-photon measurement achieves a phase

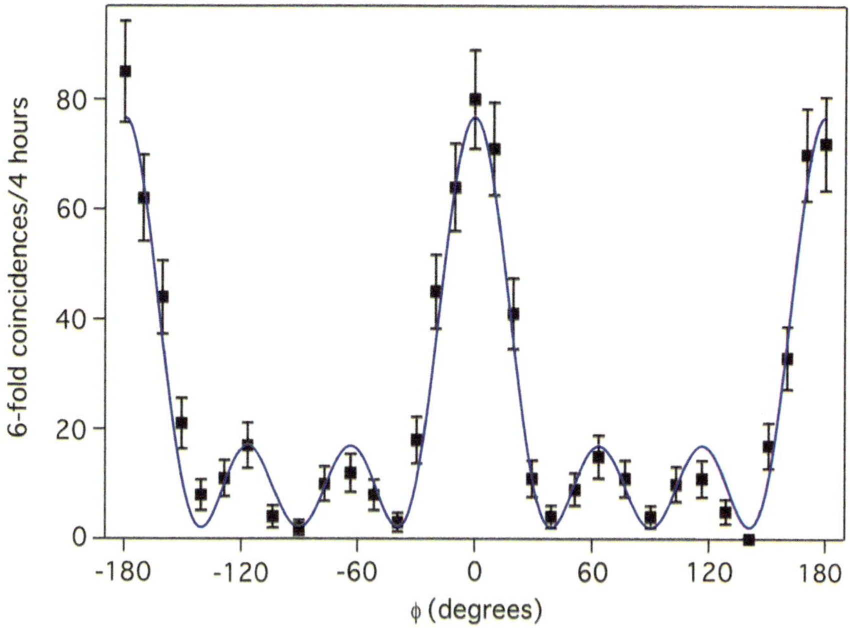

Fig. 5.5 Single-fringe measurement corresponding to the projection at the output of the interferometer, as the phase shift φ is varied. *Figure courtesy* www.nature.com

variance three times below the SNL. The results for the measurement fringe corresponding to the projection at the output of the interferometer with the variation of the phase shift φ are available in the literature and have been reproduced below (Fig. 5.5).

5.1.2.3 Electronic Speckle Pattern Interferometry (ESPI) Technique

The electronic speckle pattern interferometry (ESPI) technique is used for measuring changes in diffuse surfaces. The working of a dynamic phase-shifting electronic speckle pattern interferometer is based on producing a 9 ns pulse during each measurement and then capturing four phase-shifted interferograms during the single pulse. So, unlike the case of temporal phase-shifting interferometers, the measurements in this case do not suffer from the fringe contrast reduction and measurement errors in the presence of vibration. Also, the changes in the shape of the backplane are easily detected by making phase-shifted ESPI measurements, before and after the surface is perturbed, and then subtracting the phases. Therefore, the technique has been found to be really useful for measuring the changes in diffuse surfaces. A setup for using the spatial phase-shifting technique in the digital speckle pattern interferometry and digital shearography for non-destructive testing is shown in Fig. 5.6.

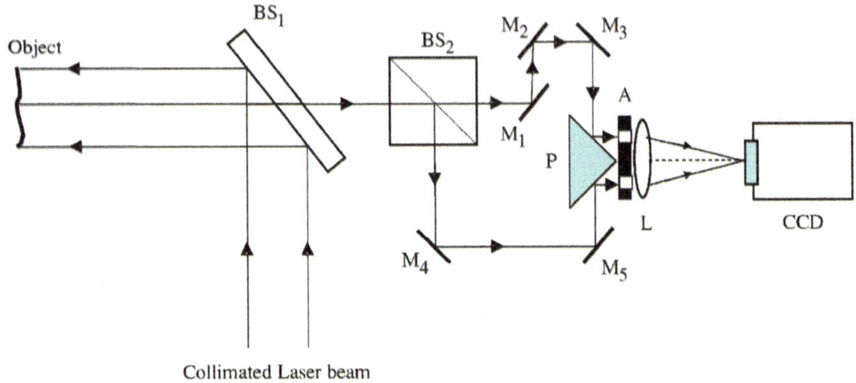

Fig. 5.6 Digital speckle pattern interferometry and digital shearography for non-destructive testing. *Figure courtesy* opticalengineering.spiedigitallibrary.org

The key components are laser, CCD, two beam splitters, and a set of mirrors, and the system works on the principle of interference.

5.1.2.4 Single-Shot Fizeau Interferometer Technique

A single-shot Fizeau interferometer technique is more difficult to make than a Twyman–Green interferometer, because of the fact that the Fizeau interferometer is more common path, and hence, it is difficult to obtain a reference and test beam with orthogonal polarization. In order to overcome this problem, a quarter-wave plate is placed between the test and reference surfaces for producing the rotation of 90 degrees in the direction of polarization of the test beam, which, however, is difficult to work for the testing of spherical optics. In some cases, the reference and test beams are tilted with respect to each other. The most suitable approach is the on-axis approach, in which a short-coherence light source is used, and the source beam consists of two time-delayed orthogonally polarized beams. By setting the path delay between the two beams equal to the path delay in the Fizeau cavity, the desired interference is achieved from the long path source beam reflected off the reference surface and the short path length source beam reflected off the test surface.

In this technique, all the beams are on-axis, and hence, it is free from the off-axis aberrations; and also because (1) both the source beams are reflected off both test and reference surfaces; and (2) only the two path-length-matched beams give interference, the fringe contrast, though reduced, is still sufficient for the testing purpose. Another advantage is that the spurious fringes are greatly reduced due to the use of a short-coherence light source. In this way, the technique has been found to be very effective in testing the optics of all shapes.

5.1.3 Optical Time-Domain Reflectometer (OTDR)

An optical time-domain reflectometer (OTDR) is a single-ended measurement tool, which provides a detailed picture of section-by-section loss, and operates by sending a high-power pulse of light down the fiber and measuring the light reflected back, thereby determining the distance (between launching point and the event) of each event by using the time taken for individual reflections to return. In this technique, the time corresponds to the acquisition duration, i.e., the period of averaging the results. Obviously, the longer acquisition times generate cleaner traces meaning long-distance traces due to the fact that as the acquisition time increases, more of the noise is averaged out, which increases (1) the signal-to-noise ratio (SNR) and (2) the OTDR's ability to detect small events. It measures/characterizes the following:

(1) Reflection and optical loss caused by every event in the link (connectors and splices), fiber ends, and detectable faults (misalignments and mismatches, dirt on connector ferrules, fiber break, and macrobends). The attenuation idea can be obtained from some of the values reported in the literature—from 0.19 dB/km at 1550 nm (0.25 dB/km for worst case) to 0.33 dB/km at 1310 nm (0.35 dB/km for worst case). OTDR in general consists of a microprocessor, a pulse trigger and generator, a laser diode, an optical coupler, a detector, an analog-to-digital converter, and a display, and the components of the device for such measurements are: data display and the control center, separated by a split bar. OTDR's basic working is on the principle—an OTDR sends short pulses (from 5 ns to 20 μs) of light into a fiber, where light scattering takes place due to the discontinuities like connectors, splices, bends, and faults, and then detects and analyzes the backscattered signals, the strength of the signal being measured for specific intervals of time, and used for characterizing the events. The pulse here corresponds to the pulse width for the test, i.e., a longer pulse allows deeper probe along the fiber, but at the cost of lesser resolution, and conversely, a shorter pulse width gives higher resolution, but at lesser distance range. The distance (d) corresponds to the distance range of the fiber span to be tested according to the selected measurement units and is computed by the following formula:

$$d = \left\{\frac{c}{n}\right\} \times \left\{\frac{t}{2}\right\}, \tag{5.1}$$

where c is the speed of light in a vacuum (2.998 × 108 m/s), t is the time delay from the launch of the pulse to the reception of the pulse, and n is the index of refraction of the fiber under test. The OTDR uses thc index of refraction (IOR) of the fiber under test to accurately calculate the distance (speed of light in fiber is different from that in the air). The flowchart of the OTDR, commonly employed and available in the literature, is reproduced below (Fig. 5.7).

It is really interesting to note that an OTDR is based on using the effects of Rayleigh scattering and Fresnel reflection for measuring and evaluating the fiber's condition. The former occurs during the passage of a pulse through the fiber and also

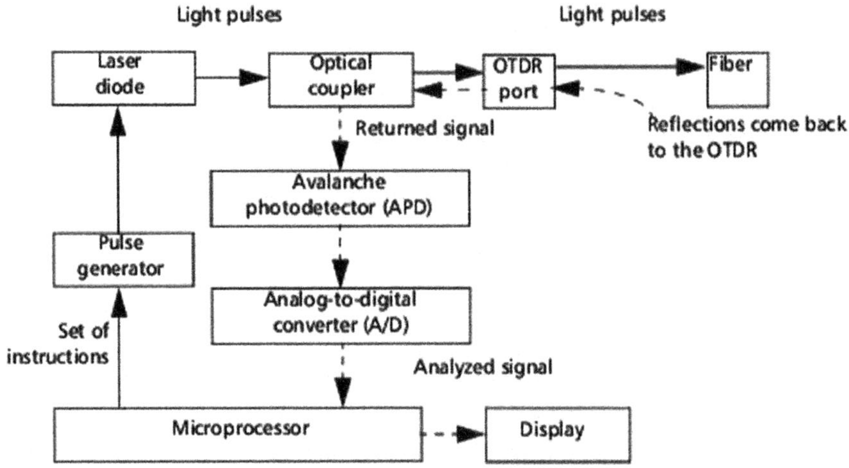

Fig. 5.7 Block diagram of the optical time-domain reflectometer (OTDR)

because of the small variations in the material like variations and discontinuities in the index of refraction, and the light is scattered in all directions. The phenomenon of backscattering refers to the small amounts of light being reflected directly back toward the transmitter. However, the problem is that the power level of the Fresnel reflection is tens of thousands of times greater than that of the backscatter. This is very simple to understand: The Fresnel reflections (the strength of the reflection depending on the degree of change in the index of refraction) take place when the light traveling through the fiber encounters abrupt changes in material density possibly due to the connections or breaks where an air gap exists. Obviously, this is very small as compared to the very large quantity of light that is reflected, as compared with the Rayleigh scattering. It is interesting to note that the OTDR uses the Rayleigh backreflections for measuring the fiber attenuation (dB/km). The backreflection level is ~75 dB. Also, the attenuation by the Rayleigh backscattering is inversely proportional to the wavelength (the higher wavelengths found to be less attenuated) by the Rayleigh backscattering. It is important to note that the Fresnel backreflections are due to the abrupt changes in the IOR (e.g., glass/air)—fiber breaks, mechanical splices, bulkheads, and connectors—and are observed as a "spike" on the OTDR trace, which appears on the screen as shown in Fig. 5.8.

The backreflection of a connector is reduced by machine polished ferrules to Super Physical Contact (SPC) polish or Ultra Physical Contact (UPC) polish. The values of UPC and APC are as: UPC reflection is ~55 dB; APC reflection is ~65 dB (as per ITU); and the Fresnel reflections are ~20,000 times higher than fiber's backscattering level. Sometimes, during the experiment, the dead zones are observed, which result from a temporary blinding of the detector following a reflective event, and in fact occur when the high return power from Fresnel reflections saturates the detector, and they are associated with every reflective event. An OTDR cannot detect or measure events in the dead zone. Thus, the dead zones affect only the reflective

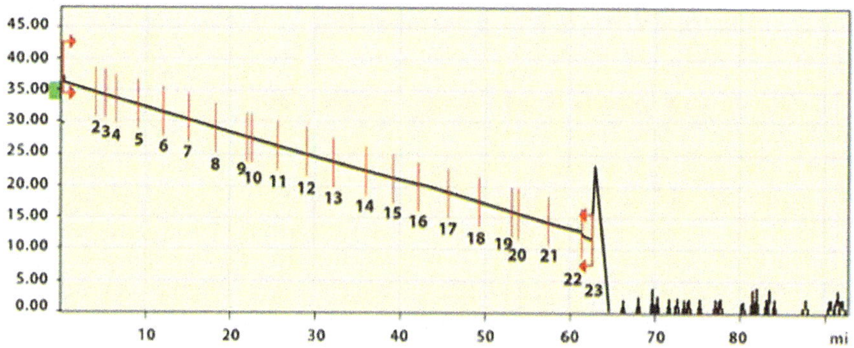

Fig. 5.8 A OTDR trace comprising tens of thousands of sampling points. *Figure courtesy* http://www.photonics.com/EDU/Handbook

events, and the event or reflective dead zone means the minimum distance between the beginning of a reflective event and the point where a consecutive reflective event should be localized. Another interesting situation is observed, when the spacing between two events is shorter than the attenuation dead zone but longer than the event dead zone, in which case, the OTDR shows the merged events. Thus, the process is quite complex and has to be handled by an experienced technologist. OTDR's ability of locating and measuring the reflectance and loss makes it useful as troubleshooting and fault locating equipment. Modern generation OTDRs are equipped with advanced functions like macrobend location, active fiber testing, and automatic trace acquisition.

5.1.4 The Fiber-Optic Measurement Technique

The fiber-optic measurement technique is used for testing a cabling infrastructure, which is important for the: (1) identification of faults or helping in trouble shooting; (2) determination of the system quality and its compliance to standard; and allowing the recording performance of the cabling at time zero. It has to be kept in mind that the testing FO cabling is an indirect process, based on the measurement of link length and loss, comparing with values calculated at design time (workmanship quality), and comparing with standard defined values (link functionality). The important components of the device are: LIGHT tracer, which is usually a red light source and launching fiber, power meter as measuring tool for light power loss, and OTDR, a graphical display of channel/link losses, location, and behavior.

5.1.5 Absolute Distance Measurements

The modern developments in laser techniques and digital image processing have made distance measurement by optical techniques very useful for many applications in industrial fields, e.g., tool calibration, aircraft industry, and robotics. These techniques are classified into two categories: (1) non-coherent methods like triangulation techniques and time-of-flight systems (measurement accuracy larger than a 1 mm) and (2) coherent methods based on interferometry, which enable high-precision measurements of distances or displacements. Classical interferometry: (i.e., one wavelength) is commonly used for high-resolution displacement measurements, with the achievable resolution better than 100 nm, though the technique has the drawback because of the incremental manner of measuring, resulting from the counting of optical fringes. Multiple wavelength interferometry (MWI) provides great flexibility in sensitivity by an appropriate choice of different wavelengths. For example, in case of two optical wavelength λ_1 and λ_2 with path difference $= L$. The phases φ_1 and φ_2 corresponding to the wavelengths λ_1 and λ_2 are given by:

$$\Delta\varphi_1 = \left\{\frac{2\pi}{\lambda_1}\right\}2L, \tag{5.2}$$

and

$$\Delta\varphi_2 = \left\{\frac{2\pi}{\lambda_2}\right\}2L. \tag{5.3}$$

Thus, the difference in these two phases is given by:

$$\Delta\varphi_{12} = (\Delta\varphi_1 - \Delta\varphi_2) = 2\pi\left[\frac{1}{\lambda_1} - \frac{1}{\lambda_2}\right]2L, \tag{5.4}$$

which can be written as:

$$\Delta\varphi_{12} = \left[\frac{2\pi}{\lambda s}\right]2L. \tag{5.5}$$

The synthetic wavelength λ_s is calculated by the equation given below:

$$\lambda_s = [\lambda_1\lambda_2(\lambda_1 - \lambda_2)]. \tag{5.6}$$

This synthetic wavelength λ_s is much longer than λ_1 or λ_2.

The range of non-ambiguity of the phase difference $\Delta\varphi_{12}$, which is also known as the synthetic phase, is therefore increased compared to the range of non-ambiguity of the classical interferometry. Moreover, the sensitivity of the measurement is reduced.

This is done by performing a number of iterations based on taking the steps: (1) estimating distance manually, (2) estimating error range and ambiguity length,

(3) adjusting tunable laser source at arbitrary λ_1 and taking the first image 1, (4) calculating λ_2: $\lambda_2 = \frac{\lambda_1}{\left[\left(\frac{\lambda_1}{\lambda_s}\right)+1\right]}$, (5) adjusting tunable laser source at this wavelength and taking the second image 2, (6) calculating $\delta\phi$ (a fraction of a fringe) using FTA, and (7) dividing the first ambiguity range by 2. In this way, more iterations are applied, and thus, the ambiguity is decreased, and error is decreased by half, which results in better accuracy. An expert and experienced technologist can do it in fewer steps.

5.2 Fabrication and Testing of Optical Components

Optical fabrication and testing has a broad spectrum of manufacturing procedures and optical testing configurations. The manufacturing of a conventional spherical lens or any other optical component starts with the generation of the optic's rough shape by grinding a glass blank, generally accomplished by using ring tool. In case of the lens surface, it is polished to its final shape. This is done by lapping, a process of rotating and rubbing the rough lens surface against a tool with the desired surface shape, with a mixture of abrasives and fluid in between. This is done both manually and automatically by machines. The skill of the technician plays a very important part, who has to choose the rotation speed and the nature of strokes (covering more of outer part or inner part). A carved (a hard material) cut or engraved to produce an object of the desired shape) pitch tool is chosen for polishing the surface of a lens. This technology is taught to the M.Tech (Applied Optics and Optoelectronics) students in the institutes like Imperial College of Science, London, UK, and Indian Institute of Technology, Delhi, INDIA, where the author learned this technology and had hands on experience of this.

The mixture of the abrasive used for polishing is called slurry, which is made from cerium or zirconium oxide in water with lubricants added for facilitating the pitch tool movement without sticking to the lens. The technician has to choose the particle size in the slurry for achieving the desired shape and finish.

5.3 Qualitative Review of Novel Studies on Optical Testing and Concluding Remarks

The field of optical testing is so important from application point of view that many studies have recently been undertaken, out of which some novel have been qualitatively discussed here. Zhang et al. [15] have established a conformal dome optical system and the related aberrations and have investigated the characteristics of the dome using Zernike aberration theory. The conformal dome has been designed with gradient index element, and the designing method along with the optimizing results has been analyzed in detail. It has been shown that the Zernike aberrations produced by the conformal dome decrease dramatically by this technique. Also, a complete conformal optical system has been designed to further illustrate the aberration correction effect of gradient index elements. It has been established that the utilization of gradient index optical elements not only improves the imaging quality, but also simplifies the structure of the conformal optical system. It is known that the diffractive optical elements can control light beyond the capabilities of traditional optical components, which offers great opportunities for numerous applications. However, modern functionalities are related to even smaller feature sizes, which cause remarkable challenges for the structure measurements in the nanoscale, and the current methods are not optimal for the characterization of diffractive structures. Optical scatterometry, however, is considered to be a good tool for non-destructive characterization, and in particular, for inline process control of the fabrication. Husu et al. [16] have discussed detailed information about a goniometric scatterometer built at the Centre for Metrology and Accreditation MIKES, basics of the inverse problem approach for solving the parameters of the structure, and first measurements for confirming the functionality and capabilities of the scatterometer.

Very recently, a conference [17] on Optical Design and Testing was held, which was dedicated to the technologies for design and testing optical surfaces, components, and systems. Papers were presented on the developments in processes, technologies, or equipment used for optical design, fabrication, or measurement. Various topics discussed were: interferometry, holography, and speckle, phase-measuring, spatial heterodyne, and static fringe analysis, absolute calibration: flats, spheres, windows, etc., measurement of aspheres, diffractive null correctors, geometric ray tests, wavefront sensors, high-spatial resolution, modulation transfer function (MTF) and encircled energy, testing in adverse environments: vibration, atmosphere, cryogenic, vacuum, etc., figure, ripple, and roughness, characterization of subsurface damage, surface profilometry: optical and scanning probe, and scatter and bidirectional reflectance distribution function (BRDF). Other important conferences on Optics and Photonics [18] and on Optics and Measurement [19] were held, respectively, in San Diego, USA, and in Liberec, Czech Republic. In addition, Bergmann and Huke [20] have written a book on Advanced Methods for optical Non-destructive Testing, which focuses on the new advancements and the progress made in the NDT technology.

It is now understood that (1) the recent advances in nanotechnologies have prompted the need for tools to accurately and non-invasively manipulate individual nano-objects, and (2) among the possible strategies, optical forces have been predicted to provide researchers with nano-optical tweezers capable of trapping a specimen and moving it in three dimensions. However, in practice, the combination of weak optical forces and photothermal issues has so far prevented their experimental realization. Berthelot et al. [21] have demonstrated the first 3D optical manipulation of single 50 nm dielectric objects with near-field nanotweezers. The nano-optical trap has been built by engineering a bowtie plasmonic aperture (bowtie shaped ridge aperture) at the extremity of a tapered metal-coated optical fiber, by performing both trapping operation and monitoring through the optical fiber, thus making these nanotweezers totally autonomous and free of bulky optical elements. It has been shown that the achieved trapping performances allow for the trapped specimen to be moved over tens of micrometers for a period of several minutes with very low in-trap intensities. It has been emphasized that this non-invasive approach is foreseen to open new horizons in nanosciences by offering an unprecedented level of control of nanosized objects, including heat-sensitive biospecimens.

Till recently, UV lithography has been used extensively for the fabrication of diffractive optical elements (DOEs), because of the advantage of this technique in that the entire structure can be written at one time, though the minimum feature size is limited to about 1 μm. Therefore, many 1D and 2D periodic grating structures may not need such fine details. However, these are essential for diffractive optics with circular structures because the spacing between features typically decreases toward the edge of the element resulting in the smallest feature falling well below 1 μm. 1D structures like sub-wavelength gratings also have smaller feature sizes throughout the structure. For such cases, advanced techniques such as focused ion beam and electron-beam lithography are required for the fabrication of finer structures. Vijayakumar et al. [22] have presented the results of DOEs fabricated with a focused ion beam system (Nova Nanolab 600 from FEI) directly on a single-mode fiber tip. It has been pointed out that the ability to write DOEs directly on fiber tip is of great importance in fields such as endoscopy and optical trapping, and that the DOE itself transforms the laser beam to a phase and intensity profile that matches the requirement, because it is located directly on the fiber, and hence has no extra alignment requirement. Vijayakumar et al. [22] have developed the most accurate method for creating the desired pattern, i.e., the DOE structure into an actually working element. Different exposure strategies for writing test structures directly with the ion beam on the fiber tip have been tested and carefully evaluated. The initial fabrication and optical test results for blazed and binary structures of 1D and circularly symmetric Fresnel axicons on optical fibers have also been presented in detail.

Huke et al. [23] have described and compared different optical NDT (ONDT) methods with respect to their characteristics and capability for different measurement tasks. ONDT may be specified in two categories, passive and active. The principles of the passive NDT just use a measurement method like view inspection, elipsometry, or reflectometry to detect defects which are easily accessible. The principles of the active NDT use an excitation force, such as heat or mechanical vibration introduced by transducers to detect hidden defects, which can be further specified into two sub-categories (1) time-/depth-resolved (phase-resolved thermography and laser ultra-sound are examples for this category) that includes measurement methods delivering detailed information of the geometric features of a hidden defect and (2) integrating that includes measurement technique (shearography, reflectometry, vibrometry, and thermography coupled with excitation method like simple heating or loading with a constant force are examples for this category) coupled with an excitation that enables detection of defects but not evaluation of their geometric features.

It has been noticed that the addition of electronics, computers, and software to interferometry has enabled enormous improvements in optical metrology. James [24] has discussed four areas in which computerized interferometric measurement improvements have been made in the measurement of surface shape and surface roughness: (1) the use of computer-generated holograms for the testing of aspheric optics, (2) phase-shifting interferometry for getting interferometric data into a computer so that the data can be analyzed, (3) computerized interference microscopes, including multiple wavelength and coherence scanning, for the precision measurement of surface microstructure, and (4) vibration-insensitive dynamic interferometers for enabling precise measurements in non-controlled environments.

Joshua and James [25] have proposed an instrument, by using a simultaneous phase sensor, which performs highly repeatable measurements over an extended range in the presence of vibration common to a laboratory setting. It has been observed that the measurement of a 4.5 μm step standard in the presence of vibration amplitudes of 40 nm produces a repeatability of 1.5 nm RMS with vertical scanning data acquired at 400 nm intervals. Also, they have demonstrated the potential of the method to tolerate larger vibration amplitudes up to or beyond a quarter wavelength and to increase the data acquisition step size to that approaching the depth of field of standard microscope imaging systems. These observations have great potential in the vibration-insensitive extended range interference microscopy. The microroughness of the optical surfaces of the optical components is very important, which has to be controlled for reducing the scattering from the components [26], and is usually done by using Nomarski Differential Contrast Microscope and Talystep Surface Profiler. Li [27] has reported the recent progress on standardization of high-reflectance testing of laser components with cavity ring-down technique and has also discussed the effect of polarization on reflectance measurement accuracy. In addition, the absorption behavior of fused silica at 193 nm has also been investigated.

Matter–wave interferometry has emerged as a powerful tool for high-precision measurements of the quantum properties of atoms, many-body phenomena and gravity. The most precise matter–wave interferometers exploit the excellent localization in momentum space and coherence of the degenerate gases. Further enhancement of the sensitivity and reduction of complexity are crucial conditions for the success and widening of their applications. Petrovic et al. [28] have introduced a multistate interferometric scheme that offers advances in both these aspects. The coherent coupling between Bose–Einstein condensates in different Zeeman states is used to generate high-harmonic output signals with an enhanced resolution and the maximum possible interferometric visibility. Also, they have demonstrated the realization of such an interferometer as a compact, easy to use, atom-chip device, which provides an alternative method for the measurement of the light–atom and surface–atom interactions and enables the application of multiparameter sensing schemes in cold-atom interferometry.

It is well known that the Passive terahertz (THz) setups require optical elements with large diameters for optimal harvesting of weak signals, and that the high f-number implies sophisticated aspheric designs to ensure optimal resolution and good energetic efficiency. The problem is that the trial-and-error testing of such optics is expensive and numerical modeling is time consuming. Siemion et al. [29] have proposed extremely cheap diffractive lenses for THz made of regular paper, which are easy to manufacture even with large diameters, and the optical function can be easily customized, that can be used for initial experimental testing of THz setups.

Characterization of the proposed diffractive lenses with time-domain spectroscopy has also been presented and discussed.

Haslinger et al. [30] have presented the experimental realization of a universal near-field interferometer built from three short-pulse single-photon ionization gratings and have observed the quantum interference of fast molecular clusters, with a composite de Broglie wavelength as small as 275 fm. It has been discussed that the optical ionization gratings are largely independent of the specific internal level structure and are therefore universally applicable to different kinds of nanoparticles, ranging from atoms to clusters, molecules, and nanospheres. It has been emphasized that the interferometer is sensitive to fringe shifts as small as a few nanometers and yet robust against velocity-dependent phase shifts, since the gratings exist only for nanoseconds and form an interferometer in the time domain. It is now well established that (1) the sub-terahertz (THz) frequency band is a noteworthy option for non-destructive testing (NDT) of non-metal aeronautics materials; (2) composite structures or laminates can be inspected for foreign objects (water or debris), delaminations, debonds, etc., using sub-THz sensors during the manufacturing process or maintenance; (3) no special security measures are needed for operation because the radiation of this frequency band is harmless to the human body; and (4) the frequency-modulated continuous-wave sensor used in this study offers a very light, compact, inexpensive, and high-performing solution. Cristofani et al. [31] have operated an automated two-dimensional scanner carrying three sensors partially covering the 70–320 GHz band, using two complementary measurement approaches: conventional focused imaging, where focusing lenses are used, and synthetic aperture (SA) or unfocused wide-beam imaging, for which lenses are not needed. It has been discussed that the conventional focused imagery offers finer spatial resolutions, but imagery is depth-limited due to the beam waist effect, whereas SA measurements allow imaging of thicker samples with depth-independent but coarser spatial resolutions. In view of these developments in this exciting field, it can be safely concluded that the field of optical testing of optical components is rapidly evolving and progressing.

References

1. V. Ronchi, Le frange di combinazioni nello studio delle superficie e dei sistemi ottici [Combination fringes in the study of surfaces and optical systems]. Rivista d'Ottica e Meccanica di precisione [J. Opt. Precis. Mech.] **2**, 9–35 (1923)
2. V. Ronchi, Due nuovi metodi per lo studio delle superficie e dei sistemi ottici (Two new methods for the study of surfaces and optical systems). Annali della Scuola Normale Superiore di Pisa - Classe di Scienze, Series **1**(15), 1–50 (1927)
3. J.A. Anderson, W. Porter Russell, Ronchi's method of optical testing. Astrophys. J. **70**, 175–181 (1929)
4. M. Mansuripur, The Ronchi test. Opt. Photonics News **8**, 42–46 (1997)
5. D.-S. Wan, D.-T. Lin, Ronchi test and a new phase reduction algorithm. Appl. Opt. **29**(22), 3255–3265 (1990)

6. S. Roux, J. Rethore, F. Hild, Digital image correlation and fracture: An advanced technique for estimating stress intensity factors of 2D and 3D cracks. J. Phys. D Appl. Phys. **42**, 214004 (2009)
7. A. Cirello, S. Pasta, Displacement measurement through digital image correlation and digital speckle pattern interferometry techniques in cold-expanded holes. Strain **46**, 581–588 (2010)
8. M.S. Kirugulige, Measurement of transient deformations using digital image correlation method and high-speed photography: application to dynamic fracture. Appl. Opt. **46**, 5083–5096 (2007)
9. L. Muravsky, A. Kmet', T. Voronyak, Two approaches to the blind phase shift extraction for two-step electronic speckle pattern interferometry. Opt. Eng. **52**, 10190 (2013)
10. J.E. Millerd, N.J. Brock, J.B. Hayes, M.B. North-Morris, M. Novak, J.C. Wyant, Pixelated phase-mask dynamic interferometer. Proc. SPIE **5531**, 304–314 (2004)
11. P. Hariharan, The Geometric Phase, in *Progress in Optics*, vol. XLVIII, ed. by E. Wolf (Elsevier, Amsterdam, 2005), pp. 149–201
12. M.N. Morris, J. Millerd, N. Brock, J. Hayes, B. Saif, Dynamic phase-shifting electronic speckle pattern interferometer. Proc. SPIE **5869**, 58691B–1 (2005)
13. P. Hariharan, *Basics of Interferometry* (Academic Press, NY, 2007)
14. G.Y. Xiang, H.F. Hofmann, G.J. Pryde, Optimal multi-photon phase sensing with a single interference fringe. Sci. Rep, **3**(Article number: 2684) (2013)
15. W. Zhang, S. Chen, C. Hao, H. Wang, B. Zuo, Z. Fan, Conformal dome aberration correction with gradient index optical elements. Opt. Express **22**, 3514–3525 (2014)
16. H. Husu, T. Saastamoinen, J. Laukkanen, S. Siitonen, J. Turunen, A. Lassila, Scatterometer for characterization of diffractive optical elements. Meas. Sci. Technol. **25**, 044019 (2014)
17. V. Conference, *Optical Design and Testing, OPTIC 2013, Optics and Photonics Taiwan, International Conference*, Taiwan, Dec 2013
18. *Conferences on Optics and Photonics*, San Diego, USA, 17 Aug 2014 → 21 Aug 2014
19. *Conference on Optics and Measurement*, Liberec, Czech Republic, 07 Oct 2014 → 10 Oct 2014
20. R. Bergmann, P. Huke, *Advanced Methods for optical Nondestructive Testing, in Optical Imaging and Metrology: Advanced Technologies* (Wiley, New Jersey, 2012), pp. 393–412
21. J. Berthelot, S.S. Aćimović, M.L. Juan, M.P. Kreuzer, J. Renger, R. Quidant, Three-dimensional manipulation with scanning near-field optical nanotweezers. Nat. Nanotechnol. **9**, 295–299 (2014)
22. A. Vijayakumar, U. Eigenthaler, K. Keskinbora, V. Pramitha, M. Hirscher, P. Spatz Joachim, S. Bhattacharya, Optimizing the fabrication of diffractive optical elements using a focused ion beam system, in *Proc. SPIE, Micro-Optics*, vol. 9130 (2014), p. 91300X
23. P. Huke, R. Klattenhoff, C. von Kopylow, R.B. Bergmann, Novel trends in optical non-destructive testing methods. J. Eur. Opt. Soc. **8**, 13043 (2013) (Rapid Publications)
24. C.W. James, Computerized interferometric surface measurements. Appl. Opt. **52**, 1 (2013)
25. T.W. Joshua, C.W. James, Vibration insensitive extended range interference microscopy. Appl. Opt. **52**, 5957 (2013)
26. K.N. Chopra, Scientific analysis of characterization techniques for optical thin films—An overview. Atti Fond G. Ronchi **70**, 465–478 (2015)
27. B. Li, Recent progress on testing of optical laser components, characterization and measurement (Th.D.), in *Optical Interference Coatings*, ***Conference Paper***, Whistler Canada, 16–21 June 2013
28. J. Petrovic, I. Herrera, P. Lombardi, F. Schäfer, F.S. Cataliotti, A multi-state interferometer on an atom chip. New J. Phys. **15**, 043002 (2013)
29. A. Siemion, A. Siemion, M. Makowski, J. Suszek, J. Bomba, A. Czerwiński, F. Garet, J.L. Coutaz, M. Sypek, Diffractive paper lens for terahertz optics. Opt. Lett. **37**, 4320–4322 (2012)
30. P. Haslinger, N. Dörre, P. Geyer, J. Rodewald, S. Nimmrichter, M. Arndt, A universal matter-wave interferometer with optical ionization gratings in the time domain. Nat. Phys. **9**, 144, 148 (2013)

31. E. Cristofani, F. Friederich, S. Wohnsiedler, C. Matheis, J. Jonuscheit, M. Vandewal, R. Beigang, Nondestructive testing potential evaluation of a terahertz frequency-modulated continuous-wave imager for composite materials inspection. Opt. Eng. **53**, 031211 (2014)

Chapter 6
Characterization Techniques for Optical Thin Films

6.1 Introduction

Characterization describes those features of the composition and structures (including defects) of a material that are significant for a particular preparation and/or study of properties sufficient for the reproduction of the material. A lot of work [1–9] has been done on evolving novel techniques with the capability of providing ever-increasing performance.

Material characterization can be classified into three categories: (1) microstructure concerned with the morphology, size minor phases, and phase distribution, (2) analytical concerned with chemical information and bonding, and (3) crystallographic concerned with crystal structure, point group, space group, and orientation. This is done by employing (1) probe of electron, ion, or photon; (2) system electronics; and (3) response to electron, ion, or photon. The various techniques and instruments employed for this purpose are: optical spectrometers, ellipsometry, X-ray diffraction, electron microscopy, XPS, SIMS, AFM, and interference (destructive and non-destructive).

Microscopy is employed to see beyond the sensitivity range and resolution limit of the eye, exploration of different wavelength regime (sensitivity) based on detectors sensitive to different regions of the EM spectrum developed and commercially available and enhancement of resolution limit of the eye (~1 arc min or 0.35 mm). This is done by using magnifying instruments like magnifying glass and optical microscope (limited by the probe wavelength), electron microscope (limited by interaction volume), and scanning probe microscope (providing atomic resolution). The importance of microscopy can be judged by the fact that the subject has fetched two Nobel Prizes (1925 for the ultramicroscope with resolution smaller than the wavelength of light, 1986 for the electron microscope with a resolving power of 500 A, and also for the STM).

Three major types of microscopes used are: (1) optical microscopes (near-field); (2) electron microscopes—scanning electron microscope (SEM), transmission electron microscope (TEM), and scanning tunneling electron microscope (STEM); and

K. N. Chopra, *Optoelectronic Gyroscopes*, Progress in Optical Science and Photonics 11,
https://doi.org/10.1007/978-981-15-8380-3_6

(3) scanning probe microscopes—atomic force microscope (AFM) and scanning tunneling microscope (STM).

6.2 Structural Features—Magnifications

The structural features of interest are of four types with the respective magnifications of two orders of magnitude higher than the previous one. These are given below:

1. Macrostructure: The common experimental methods are: visual inspection, X-ray radiography, and ultrasonic inspection. The characteristic features are: production defects, porosity, and cracks. The magnification in this case is unity.
2. Mesostructure: The common experimental methods are: optical microscopy and scanning electron microscopy. The characteristic features are: grain and particle size, and morphology. The common experimental methods are: optical microscopy and scanning electron microscopy. The magnification in this case is 100.
3. Microstructure: The common experimental methods are: scanning and transmission electron microscopy and atomic force microscopy. The characteristic features are: dislocation substructures; grain and phase boundaries, and precipitation. The magnification in this case is 10,000.
4. Nanostructure: The common experimental methods are: X-ray diffraction, SPM, and HRTEM. The characteristic features are: crystal and interface atomic structure, defects, nanoparticles and their structures. The magnification in this case is 1,000,000.

6.2.1 Comparison of Advanced Microscopes with Optical Microscopes (OMs)

The visible light wavelength is in the range: 4000–7000 Å; and the image formation follows the laws of physical optics, resulting in the resolution limited by the diffraction limit and wavelength of the light. Electron microscopy makes use of the characteristic of the wave nature of high-speed electrons. The electrons accelerated to 10,000 eV have a wavelength of ~0.12 Å, and hence, the electron microscope provides much higher diffraction limited resolution. The physical nature of a solid surface is studied by the scanning electron microscopy (SEM), and the chemical composition of surface is evaluated by the analytical electron microscopy (AEM).

6.3 Characterization Techniques

There are many types of characterization techniques, for studying different aspects of optical thin films. These are based on different concepts and are technically analyzed below.

6.3.1 *Analytical Characterization*

Analytical characterization provides the chemical information about the atomic species involved in their distribution, purity, and bonding techniques. The chemical information and purity are done by X-ray analysis–energy-dispersive X-ray spectroscopy (EDS) and X-ray fluorescence (XRF). EDS is based on making use of the X-ray spectrum emitted by a solid sample bombarded with a focused beam of electrons to obtain a localized chemical analysis. XRF is based on the emission of characteristic secondary or fluorescent *X-rays* from a material, which has been excited by bombarding with high-energy X-rays or *gamma rays*. The optical analysis is done by Fourier transform infrared (FTIR) technique; and the related absorption and emission spectra. The bonding is done by X-ray photoelectron spectroscopy or electron spectroscopy for chemical analysis. The EDS line profile software automation allows simplified composition profiling at nanometer resolution using EDS, e.g., Si80Ge20 islands grown on Si at high temperature ~800 °C followed by Si capping. In the same way, the EDS mapping allows visualization of the phase separation process, which can be coupled with point-by-point quantitative analysis, e.g., $Cu_{50}Ag_{50}$ ball milled at 230 °C.

6.3.2 *X-Ray Photoelectron Spectroscopy (XPS)*

Photoelectron spectroscopy is based on utilizing photoionization and analysis of the kinetic energy distribution of the emitted photoelectrons, thereby evaluating the composition and electronic state of the surface region of a sample. With the advent of the development of synchrotron radiation sources, it has become possible to undertake the high-resolution studies with radiation spanning a much wider and more complete energy range (5–5000 + eV). However, not many studies have been carried out on this topic due to the expensive instrumentation, complexity of the process, and also very limited availability of such sources.

The physical principle involved is simple: The photoelectron spectroscopy is based on a single photon in/electron out process and the underlying process is comparatively simpler than the Auger process. The technique uses the monochromatic sources of radiation, i.e., photons of fixed energy. The kinetic energy distribution of the emitted photoelectrons, implying the number of emitted photoelectrons as a function of their kinetic energy is measured by using a suitable electron energy analyzer, thus recording a photoelectron spectrum. This technique has been observed to be necessarily surface sensitive.

The XPS is based on the simple principle: The photon is absorbed by an atom in a molecule or solid, leading to ionization and the emission of a core (inner shell) electron. Hence, the kinetic energy distribution of the emitted photoelectrons can be measured using any appropriate electron energy analyzer, thus recording a photoelectron spectrum. The calculations are based on the following equation:

$$\mathrm{KE} = h\upsilon - \mathrm{BE} \tag{6.1}$$

where BE is the binding energy.

Commonly employed X-ray sources are those giving rise to: Mg Ka radiation: $h\upsilon = 1253.6$ eV; and Al Ka radiation: $h\upsilon = 1486.6$ eV. The process is illustrated in Fig. 6.1.

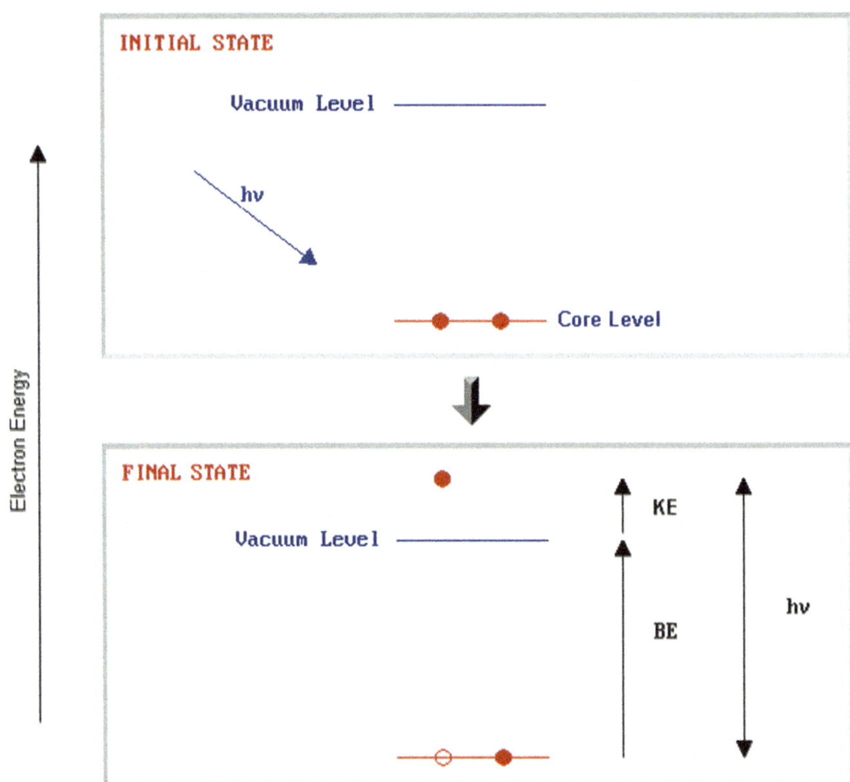

Fig. 6.1 Absorption of the by an atom in a molecule or solid (top), leading to the ionization and the emission of a core (inner shell) electron (bottom) in XPS. *Figure courtesy* http://www.chem.qmul.ac.uk/surfaces/scc/scat5_3.htm

Obviously, each element gives a characteristic set of peaks in the XPS spectrum. The BE of each peak is specific to the emitting element and therefore can be used to identify the element. Also, the intensity of the peak and the area can be used for the quantitative estimation of the element. However, it has to be noted that the exact binding energy of an electron, in addition depends upon: (1) the formal oxidation state of the atom; and (2) the local chemical and physical environments. Changes in either of these give rise to small shifts in the peak positions in the spectrum, the so-called chemical shifts.

6.3.3 Structure by X-Ray Diffraction

X-ray diffraction is sensitive to thin films of atomic dimensions to thicknesses of many tens of microns, due to the X-ray wavelengths used and the very high diffraction space resolutions achievable. X-ray methods are mostly non-destructive, in that the sample preparation is not required, and they can provide a very appropriate analysis to obtain structural information of dielectric thin films and multilayers. Analysis can be performed for whole spectrum of material types from perfect single crystals to amorphous materials, and the choice of the X-ray diffraction analysis procedure depends mainly on the quality of the structural form.

X-ray diffraction is based on the constructive interference of monochromatic X-rays and a thin films coated sample. The X-rays are generated by a cathode ray tube, filtered to produce monochromatic radiation, collimated to concentrate, and finally directed toward the coated sample.

The interaction of the incident rays with the sample produces constructive interference when conditions satisfy Bragg's law:

$$n\lambda = 2d\,\sin\theta, \tag{6.2}$$

which relates the wavelength of electromagnetic radiation to the diffraction angle, and the lattice spacing in the sample. Subsequently, these diffracted X-rays are detected, processed, and counted. Finally, the material is identified by comparing the diffraction peaks to d-spacings because each material has a set of unique d-spacings. Typically, this is done by comparing the d-spacings with standard reference patterns.

6.3.4 Mass Spectrometry

Mass analyzers are commonly used for secondary ion mass spectrometry (SIMS), which is the technique used to analyze the composition of solid surfaces and *thin films* by *sputtering* the surface of the specimen with a focused primary *ion beam* and collecting and analyzing the ejected secondary ions. This is done by measuring the mass/charge ratios of these secondary ions with a *mass spectrometer* to determine the elemental, isotopic, or molecular composition of the surface to a depth of 1–2 nm. The magnetic sector analyzers are based on very simple principle: In the magnetic field, the ion experiences a centripetal force of Bev, which has to be balanced by the centrifugal force of the ion $\frac{mv^2}{r_m}$.

This implies that

$$F_B = Bev = F_C = \frac{mv^2}{r_m} \tag{6.3}$$

The *m*/*e* separation is achieved by varying either *B* (magnetic field) or r_m (radius of curvature). The detector position is kept fixed, thereby fixing r and hence *m*/*e* separation is achieved by varying *B*. The schematic of the magnetic sector analyzer is shown below:

As shown in Fig. 6.2 the modern mass spectrometers have non-normal pole faces for the entrance and exit of the ion beam to the magnetic sector, and the fringing fields in this configuration result in compressing the ion beam in vertical direction, both in and out of the screen, during its passage through the sector. The slits are arranged at the ion beam crossovers for the cleanest separation, meaning the highest mass resolution with similar *m*/*z* values. The right part of the beam represents the ions with higher *m*/*z* values, which in fact do not pass through the spectrometer. Interestingly, a suitable combination of a magnetic sector and an electrostatic sector provides a double focusing device.

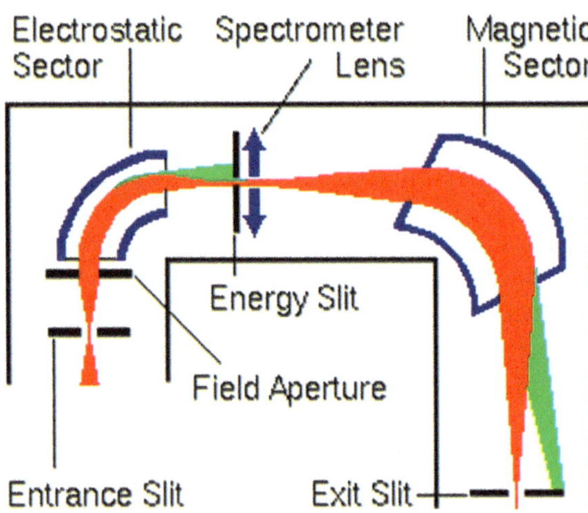

Fig. 6.2 Magnetic sector analyzer. *Figure courtesy* www.eag.com

6.3.5 Scanning Electron Microscope (SEM)

SEM has been very useful in evaluating the carbon nanotubes (CNTs). The upcoming field of analysis of CNTs is done for showing the density, alignment, and height, which are used for evaluating the CNTs. In addition, it throws some light on diameter of the nanotubes. The transmission electron microscopy (TEM) is an electron diffraction technique, which samples the bulk of the sample due to the geometry of the system. The TEM images also give very high resolution and hence are used in studying the structure of optical thin films

6.3.6 Scanning Tunneling Microscope (STM)

A scanning tunneling microscope (STM) is an instrument for imaging surfaces at the atomic level. For an STM, good resolution is considered to be 0.1-nm-lateral resolution and 0.01-nm-depth resolution. With this resolution, individual atoms within materials are routinely imaged and manipulated. The STM can be used not only in ultrahigh vacuum but also in air, water, and various other liquid or gas ambient, and at temperatures ranging from near zero Kelvin to a few hundred degrees Celsius.

The STM is based on the concept of quantum tunneling, which explains that when a conducting tip is brought very near to the surface to be examined, a bias (voltage difference) applied between the two can allow electrons to tunnel through the vacuum between them, and the resulting tunneling current is a function of the tip position, applied voltage, and the local density of states (LDOSs) of the sample. Information is acquired by monitoring the current as the tip's position scans across the surface and is usually displayed in image form. STM is really a challenging technique, as it requires extremely clean and stable surfaces, sharp tips, excellent vibration control, and sophisticated electronics, but is at present commercially available. The schematic of an STM is shown in Fig. 6.3.

6.3.7 Reflection High-Energy Electron Diffraction (RHEED)

RHEED is a technique useful for characterizing the surface of materials. RHEED systems gather information only from the surface layer of the sample, which separates it from other materials characterization techniques that also rely on the diffraction of high-energy electrons. A simple RHEED system requires an electron source (gun), photoluminescent detector screen, and a sample with a clean surface. However, the modern RHEED systems have additional parts for optimizing the technique.

The electron gun produces a beam of electrons which strike the sample at a very small angle of incidence on the sample surface. Incident electrons diffract from atoms at the surface of the sample, and a small fraction of the diffracted electrons

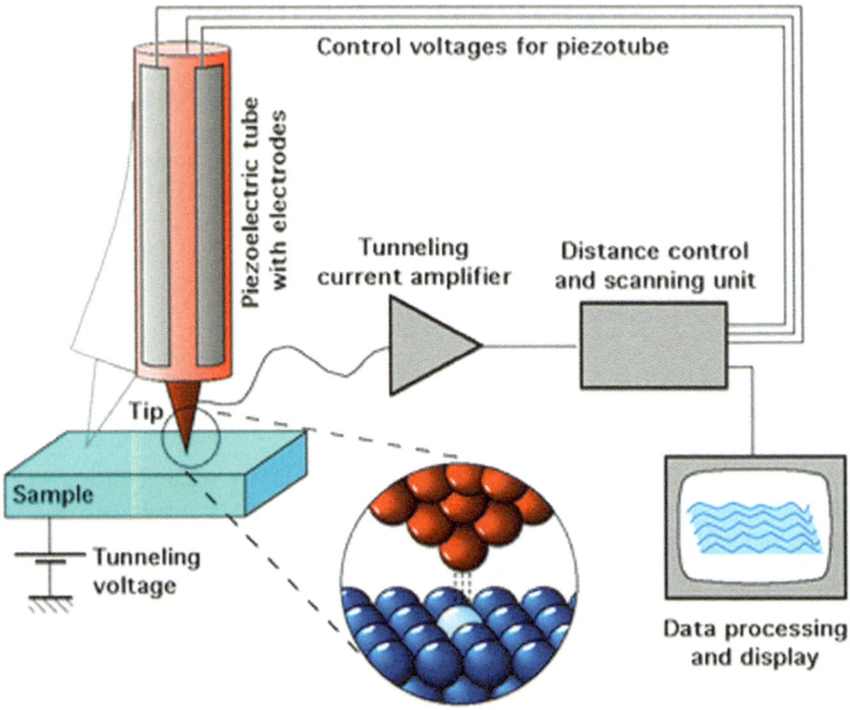

Fig. 6.3 Block diagram of the scanning tunneling microscope. *Figure courtesy* www.physicscentral.com

interfere constructively at specific angles, and hence lead to the formation of the regular patterns on the detector. The electrons interfere according to the position of atoms on the thin film coating on the sample surface, and hence, the diffraction pattern at the detector is a function of the coating surface. In this way, the technique is really useful for ensuring the smoothness of the coatings, hence minimizing the scattering losses. The basic setup of a RHEED system is shown in Fig. 6.4.

The electrons follow the path indicated by the arrow and approach the sample at angle θ. The diffraction of the electrons takes place from the sample surface, and some of these diffracted electrons reach the detector and form the RHEED pattern. Also, the reflected beam follows the path from the sample to the detector. Low-energy electron diffraction (LEED) is also surface sensitive, but it achieves surface sensitivity through the use of low energy electrons.

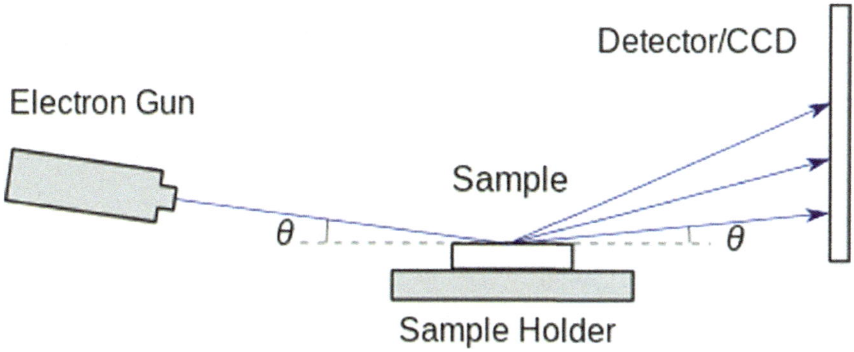

Fig. 6.4 Block diagram of the reflection high-energy electron diffraction. *Figure courtesy* en.wikipedia.org

6.3.8 *Atomic Force Microscopy (AFM)*

Atomic force microscopy (AFM), also known or scanning force microscopy (SFM), is a very high-resolution type of scanning probe microscopy, with resolution ~ fractions of a nanometer, i.e., more than 1000 times better than the optical diffraction limit. The AFM is a very important tool for imaging, measuring, and manipulating thin films at the nanoscale level. Interestingly, the information is obtained by feeling the surface with a mechanical probe.

The principle is simple—the piezoelectric elements that facilitate tiny but accurate and precise movements on (electronic) command are used for very precise scanning. Also, in case of the advanced versions, currents can be passed through the tip to probe the electrical conductivity or transport of the underlying surface. However, this is much more challenging and is currently being used by few research groups reporting consistent data. The block diagram of atomic force microscope using beam deflection detection is shown in Fig. 6.5.

It has to be understood that as the cantilever is displaced because of its interaction with the surface, the reflection of the laser beam is also be displaced on the surface of the photodiode in the same manner.

Some of the figures of SiO_2 coatings and microroughness of TiO_2 coatings taken on AFM have been shown in Figs. 6.6 and 6.7.

6.3.9 *Fourier Transform Infrared (FTIR) Technique*

Fourier transform infrared (FTIR) spectroscopy is a technique which is used to obtain an infrared spectrum of absorption, emission, photoconductivity or Raman scattering of a thin film. Fourier transform infrared spectroscopy is a combination of IR absorption spectroscopy and a Michelson interferometer, based on recording the absorption

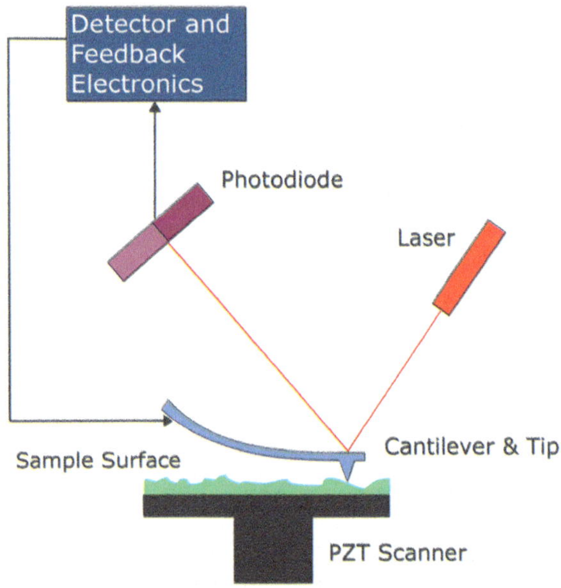

Fig. 6.5 Block diagram of the atomic force microscope. *Figure courtesy* en.wikipedia.org

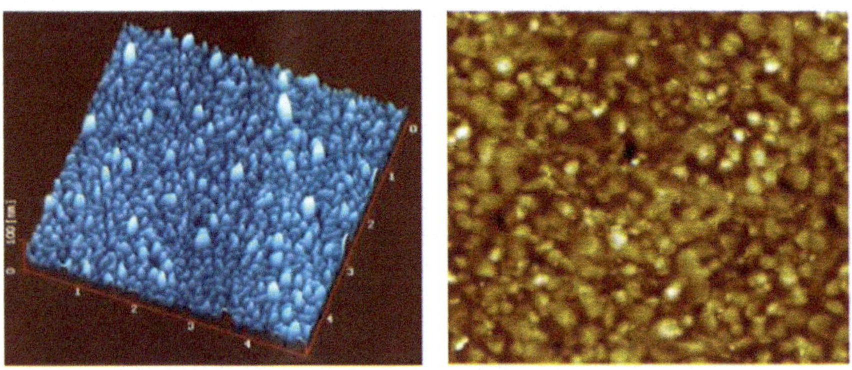

Fig. 6.6 Figures of SiO_2 coatings taken on AFM. *Figure courtesy* https://www.laserfocusworld.com

as a function of time, according to the movement of the mirror in one of the interferometer arms. Then, the resulting interferogram is Fourier transformed into the frequency space to obtain an absorption spectrum. Interestingly, the spectrum is sensitive to the vibrations of the molecular bonds of the sample material.

An FTIR spectrometer simultaneously collects spectral data in a wide spectral range. This has a great advantage over a dispersive spectrometer which measures intensity over a narrow range of wavelengths at a time. The term Fourier transform infrared spectroscopy originates from the fact that a Fourier transform is required to

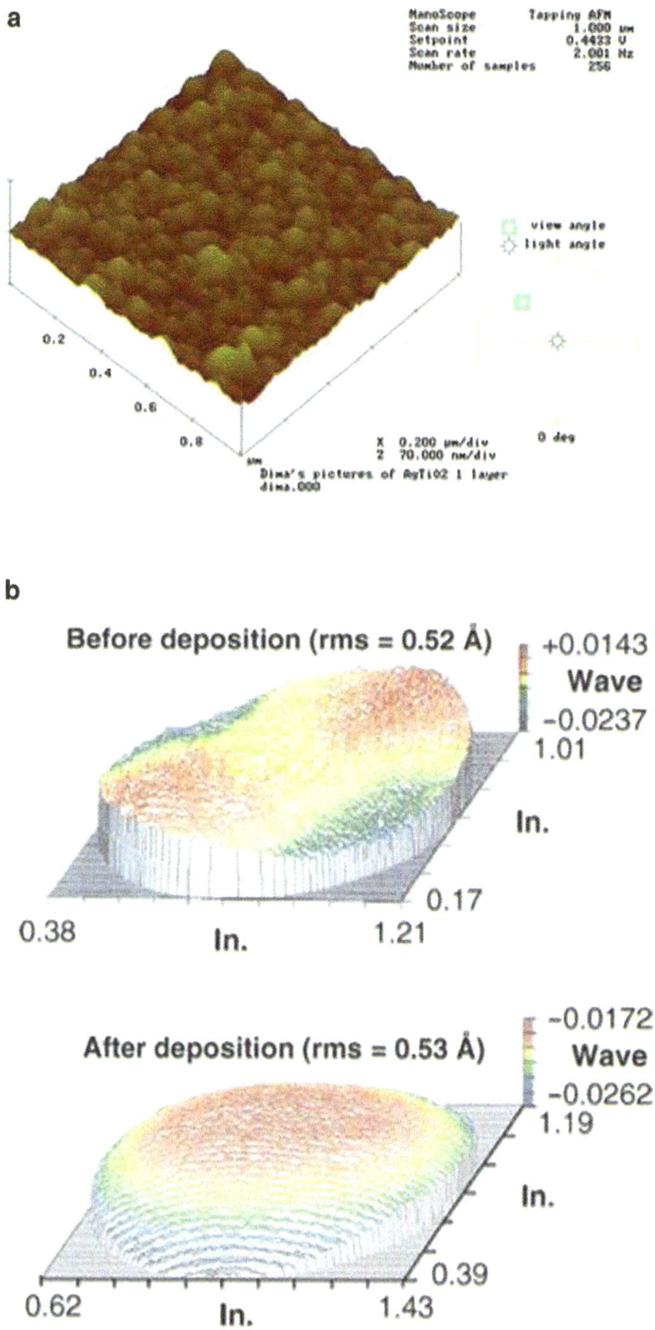

Fig. 6.7 **a** Nanoscope of TiO_2 coatings. **b** Figures of microroughness of TiO_2 coatings before and after depositions. *Figure courtesy* https://www.laserfocusworld.com

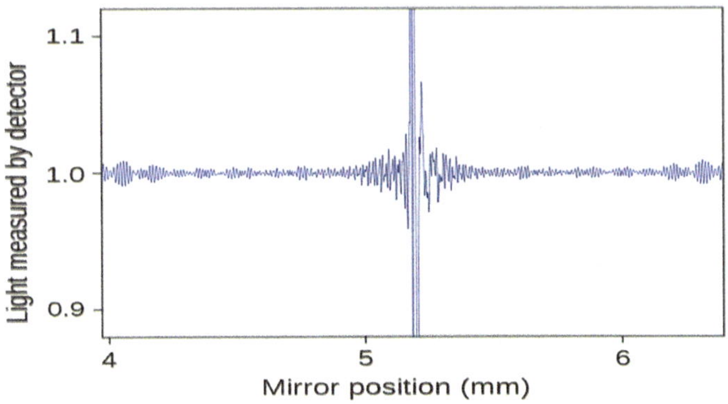

Fig. 6.8 FTIR interferogram. *Figure courtesy* en.wikipedia.org

convert the raw data into the actual spectrum. An FTIR interferogram is shown in Fig. 6.8.

It may be noted that the central peak is at the zero path difference (ZPD) position or the zero retardation where the maximum amount of light passes through the interferometer to the detector. It has to be noted that the interferogram is in the length domain, and the Fourier transform (FT) inverts the dimension, so the FT of the interferogram is in the reciprocal length domain, i.e., the wavenumber domain. Thus, it is obvious that the spectral resolution (in wavenumbers per cm) is equal to the reciprocal of the maximum retardation in cm.

To have an idea, the calculations show that a 4 cm^{-1} resolution is obtained if the maximum retardation is 0.25 cm, in case of the simple FTIR instruments. Much higher resolution can be obtained by increasing the maximum retardation. The technique is used to find the presence of any unwanted groups (Si–O, OH, etc.).

FTIR spectra are also used for studying the effect of annealing at different temperatures of doped thin, e.g., ZrO_2-doped TiO_2 films. Interesting results are available in the literature, which have been reproduced in Fig. 6.9.

Figure 6.10 shows the infrared absorption spectrum of the 5% ZrO_2-doped TiO_2 films annealed at different temperature. The peak at 2360 cm^{-1} is explained to result from the adsorbed H_2O molecules, which are not removed completely after sol-gel coating. The peaks at 1242, 1111, 1035, and 860 cm^{-1} are interpreted to correspond to the vibration mode of Ti–OH. The band around 665 cm^{-1} is especially interesting and is attributed to the vibration mode of Ti–O–Ti bond, whereas the band around 455 cm^{-1} is considered to be the O–Ti–O band corresponding to the crystalline titania in the anatase form, which is one of the three mineral forms of titanium dioxide, the other two being brookite and rutile. It is observed that the intensity in vibration bands located in the vicinity of 665 and 455 cm^{-1} increase on increasing the annealing temperature, which clearly indicates that the number of Ti–O–Ti and O–Ti–O links of titanium dioxide crystallization is also growing.

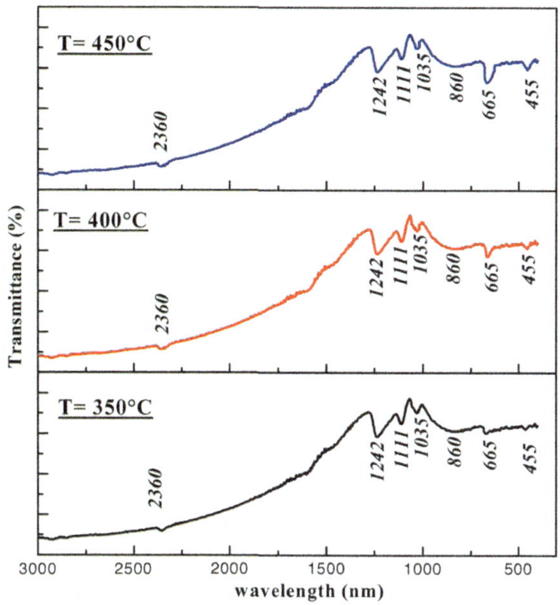

Fig. 6.9 Infrared absorption spectrum of the 5% ZrO_2-doped TiO_2 films annealed at different temperature. *Figure courtesy* www.intechopen.com

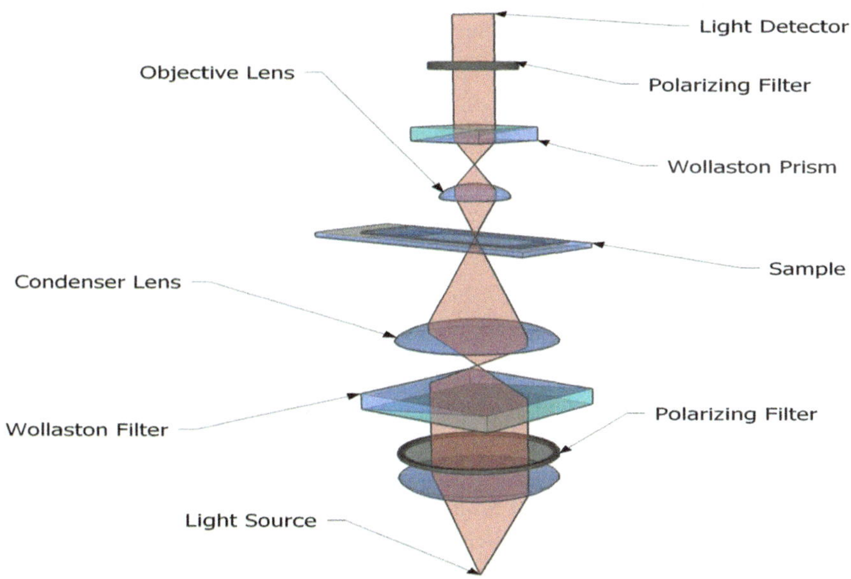

Fig. 6.10 Block diagram of the differential interference contrast microscope. *Figure courtesy* en.wikipedia.org

6.3.10 Differential Interference Contrast (DIC) Microscopy

Differential interference contrast microscopy, also known as Nomarski interference contrast (NIC) microscopy or simply as Nomarski microscopy, is indeed an important optical microscopy illumination technique used for enhancing the contrast in unstained, transparent samples. DIC is based on the principle of interferometry to gain information about the optical path length of the sample, and thus to see features, which are otherwise invisible. In this system, a relatively complex lighting scheme produces an image with the object appearing black to white on a gray background, and the image is similar to that obtained by phase-contrast microscopy with the added advantage that it is without the bright diffraction halo.

The technique is based on separating a polarized light source into two orthogonally polarized mutually coherent parts which are spatially displaced, i.e., sheared at the sample plane, and recombined before taking the observation. It has to be understood that the interference of the two parts at recombination is sensitive to their optical path difference (the product of refractive index and geometric path length). By adding an adjustable offset phase, the interference at zero optical path difference in the sample is determined, and this contrast is proportional to the path length gradient along the shear direction, which gives the appearance of a three-dimensional physical relief corresponding to the variation of optical density of the sample, and thus emphases lines and edges, though not providing a topographically accurate image. Thus, the technique is extremely useful in studying the scratches and digs in the surface and hence the microroughness of the uncoated substrates and the coated samples.

The components of the basic differential interference contrast microscope setup are shown below.

6.3.11 Spectrophotometers

A spectrophotometer is an important instrument for characterizing the RLG mirrors, which has a monochromator, for producing a light beam having wavelengths in a narrow band around the desired selected wavelength, and a component for measuring the ratio of that beam's intensity as it enters and leaves a cuvette. Though the spectrophotometers and photometers are limited in their ability to measure highly absorptive samples, they are very useful for the measurement of dielectric coatings of low absorption loss and give a good idea about the loss though for the most accurate results, cavity loss meter has to be used. Still, to have an idea about their performance for studying the highly absorbing samples, let us consider the following case.

For the case $A = 3$, only one-thousandth of the light incident on the sample goes into the detector, and, in most instruments, even the detector's self-noise begins to compete with the signal produced by that light. So for the case of the higher absorption, the detector noise completely dominates the light-induced response, and therefore, the instrument reports just the noise, and not the light passing through the sample. In addition to noise response, there is always some stray light within the instrument, and so the detector responds to this also, leading to erroneous results. An ideal and perfect detector must produce a response in which the reported absorption, A (rep), equals the actual absorption A, i.e.,

$$A(\text{rep}) = A \tag{6.4}$$

For the case of absorption beyond this limit of this dynamic range, the response may be computed according to the following expression:

$$A(\text{rep}) = \left(\mathrm{e}^{-\alpha A}\right)A, \tag{6.5}$$

in which the proportionality factor, $\mathrm{e}^{-\alpha A}$, becomes less than 1, as A increases. Also, α is a small number.

It is important to note that while analyzing the darker beers and worts, the designer has to study the samples having spectra which, for certain wavelengths, exceed instrument's dynamic range, and so correction based on experience and knowledge of the designer, has to be applied, which is quite difficult. Some results are available in the literature [10] and have been reproduced below.

In this Fig. 6.11 the spectrum labeled "1 cm no dilution" is an example of such a spectrum. As can be observed, portions of that spectrum at wavelengths shorter than 400 nm are dominated by noise and stray light. So the designer has to be well aware of the limitations of his instrument, and when absorptions approach the limits of instrument linearity, readings have to be rejected, since applying correction is really a complicated problem.

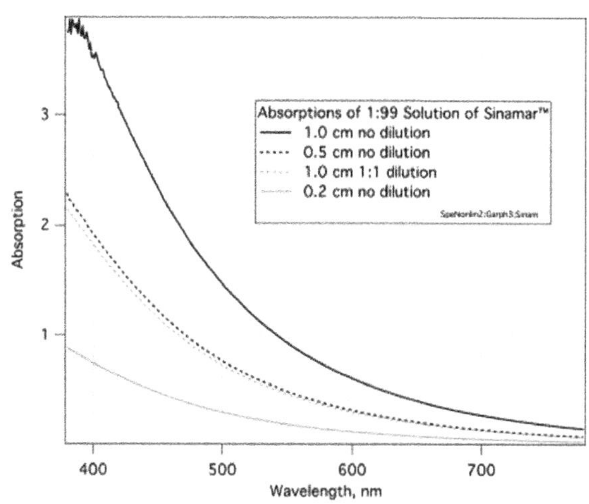

Fig. 6.11 Measured absorption spectra of sinamar in various paths and diluted. *Figure courtesy* de Lange [10]

6.3.12 UV-VIS Spectrophotometers

Spectrophotometers (UV-VIS) are the instruments for absorption spectroscopy. These are very useful for analyzing the RLG coatings. In fact, an absorption spectrophotometer is a device used for measuring absorbed light intensity as a function of wavelength. In UV-visible spectrophotometers, a beam of light from a suitable UV and/or visible light source is made to pass through a prism or diffraction grating monochromator, to produce a parallel beam of light, which is then made to pass through the sample to be analyzed before falling on the detector. The main components of a UV-visible spectrophotometers are: the light source, monochromator, sample holder, detector, and interpreter. The standard light source commonly employed consists of a deuterium arc (190–330 nm) and a tungsten filament lamp (330–800 nm), which produces a light beam in the 190–800 nm spectral region and hence provides the UV'-VIS spectrum. The monochromator produces a compact parallel optical path and minimizes the optical aberrations. Modern instruments use grating monochromators in reflection mode as the dispersing element, for providing an improved performance. The experimental setup of an optical absorption spectrophotometer [11] is shown below.

As shown, there are two types of spectrophotometers: single and double beam. In single beam setup, the components are in single beam sequence; thus, this instrument is less expensive to design and make; and also quite easy to use and maintain. On the other hand, a single beam spectrometric instrument requires a reference sample to be measured separately from the test sample. In this spectrophotometer, the light from the source is split into two separate beams after passing through the monochromator (Fig. 6.12b). Whereas one beam is used for the sample, the other is used for reference purpose. Clearly, this configuration is advantageous because sample and reference readings can be taken simultaneously, and so the measurement becomes independent from variations in the intensity and spectral composition of the light source. It is important to note that a single beam spectrophotometer should always be switched on some time before use to allow the lamp to reach a constant temperature. Also, some double beam spectrophotometers use a beam chopper, in which by blocking one beam at a time, measurement of sample and reference beams by a single photomultiplier is done at the same time.

As we have discussed earlier that scattering from the RLG mirrors affects the performance of the RLG, and hence its measurement is very important for minimizing it. Spectroscopy and Optics: Laser mirrors: It has been observed that the high reflectance is measured best with the cavity ring-down spectroscopy. Also, for mirrors with reflectance values more than 99.99%, measuring total loss through cavity ring-down spectroscopy (CRDS) is much more accurate and reliable than the conventional methods used for ordinary laser mirrors. IAN STEVENSON and CORY BOONE of Edmund Optics, Barrington, NJ, have discussed the reflectance measurement of high-reflectance mirrors, with reflectance values ranging from 99.8 up to 99.999%, which are critical components in most of the laser systems, especially RLGs for beam steering for maximizing throughput. Mostly, the mirror reflectance is

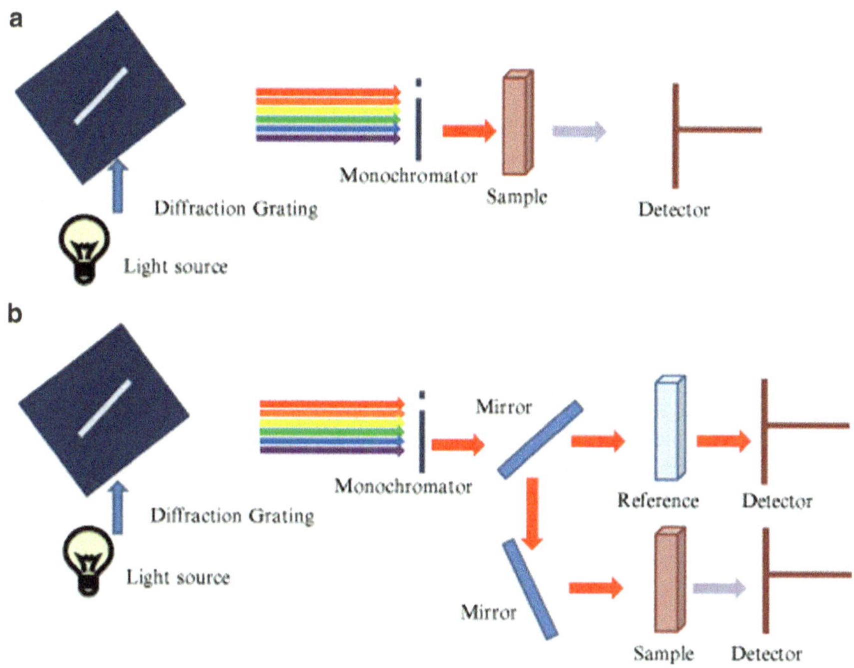

Fig. 6.12 Schematic of an optical absorption spectrophotometer. **a** Single beam spectrophotometer and **b** double beam spectrophotometer. *Figure courtesy* Nilapwar et al. [11]

evaluated by measuring the transmission by using spectrophotometers, by assuming that the remaining part of the light is reflected. This, however, is a wrong assumption since it does not consider the scattering and absorption losses and thus leads to forming an opinion of much higher but false reflectance values. In such cases, a more accurate method of determining reflectance is by measure the total loss by cavity ring-down spectroscopy (CRDS). Minimal losses and high-reflectance values of the RLG mirrors are crucial for maximizing throughput. Dielectric coatings are used for maximizing the reflectance and laser-induced damage threshold (LIDT), >20 J/cm^2 for 20 ns, 1064 nm pulses. In RLGS, for maximizing throughput reflectance values up to 99.999% may be required. These ultrahigh reflectance mirror coatings have to be deposited by the tightly controlled processes like double ion beam sputtering (DIBS). The fact is that any light not reflected is either transmitted through the system, or absorbed, or scattered. The undesired transmission leads to a beam focusing at some point in the system, which leads to safety problem or laser-induced damage. Also, the absorption my lead to laser-induced damage through thermal effects. Though the scattered light is not well-focused and is less in intensity than the original beam, it results in reducing throughput and introducing noise into the system and standard method of measuring reflectance. Thus, it can be easily visualized that the direct measurements of reflectance with a spectrophotometer have an uncertainty of ±0.5%, which can be brought down to ±0.1%, by a skilled worker by taking all precautions; though this is still not accurate enough for mirrors with reflectance higher than 99.9%, where the difference between success and failure can be even <0.1%. This problem can be overcome only by using cavity ring-down spectroscopy(CRDS), which provides an ideal solution for determining the reflectance of highly reflective optics by measuring the total loss of the optic, which includes absorption, transmission, and scatter from which the reflectance of the optic can be determined by subtracting the measured loss from 100%.

The CRDS technique is based on forming a resonant cavity by two highly reflective mirrors with reflectance values nearly 99.99%. Then, a laser pulse is made to pass through the cavity, in which it oscillates between the mirrors, losing a small amount of intensity at each reflection due to transmission, absorption, and scattering. The decreasing intensity of the reflected light is reorder by an intensity detector placed after one of the mirrors, as a function of time. In this way, the total loss of both mirrors is determined by the decay time, or "ring down," of the reflected light inside the cavity. Obviously, the sensitivity of directly measuring a laser mirror's reflectance with an uncertainty of ±0.1% is two orders of magnitude greater than directly measuring the mirror's loss with an uncertainty of ±10% through CRDS.

The intensity of the laser pulse inside of the resonant cavity is described by:

$$I(t) = I_0 \mathrm{e}^{-Ttc/2L}, \tag{6.6}$$

where $I(t)$ is the intensity of the laser pulse as a function of time, I_0 is the initial laser pulse intensity, T is the total cavity mirror loss, t is the time, c is the speed of light, and L is the length of the cavity. Since the loss value measured in a CRDS system is the loss of both the mirrors in the cavity, three tests have to be performed using reference mirrors to determine the loss of one unknown mirror. It may be noted that two known reference mirrors are used to form the resonant cavity for the calibration purpose. It is easy to note that the cleanliness is critical to accuracy with a CRDS, since any dust inside the instrument cavity or contamination on the mirror surfaces results in an increased loss and reduced reflectance. Therefore, CRDS systems are always operated in a clean room. Clearly, the direct measurement of loss in CRDS is ideal for characterizing highly reflective laser optics like RLG mirrors since it is much easier to accurately measure a small amount of loss than a large reflectance value. The reflectance values of test component compare around 99.9–99.999%, resulting in an uncertainty <0.001% of the total reflectance, i.e., a sensitivity of 1×10^{-5}. This is much better than an accuracy of around 0.1%, i.e., 1×10^{-3} sensitivity.

6.3.13 Photothermal Spectroscopy

Photothermal spectroscopy is a group of various high sensitivity spectroscopy techniques, followed in laboratories for evaluating optical absorption and thermal characteristics of a sample, e.g., a multilayer-coated sample like RLG mirror. These techniques are based on the change in the thermal state of the sample due to the absorption of radiation. The working is simple: the light absorbed by emission results in heating of the sample, which thereby raises temperature and hence influences the thermodynamic properties of the sample like the dielectric-coated RLG mirror placed adjacent to it. Measurement of the effects in the form of changes in temperature, pressure, or density occurring due to the optical absorption, form the basis for the photothermal spectroscopic measurements.

In analogy to photoacoustic spectroscopy, photothermal spectroscopy is an indirect technique for measuring optical absorption, since it is not involved in the direct measurement of the light resulting in the absorption. However, it is similar to the photoacoustic technique, since both directly measure the absorption, rather than calculating it from the transmission, as in the case of the transmission spectroscopy.

Thus, we can understand that this technique is highly sensitive, since in the transmission techniques, the absorbance has to be calculated in terms of the difference between total light impinging on the sample and the transmitted light, thus suffering from the problems of accuracy while dealing with small differences between large numbers, and also due to scattering and reflection contributing to some error. However, the advantage in case of the photothermal spectroscopies is that the absorption is directly proportional to the absorption and is zero when there is zero true absorption, in spite of contribution of reflection or scattering.

6.3.14 Photothermal Deflection Spectroscopy (PDS)

PDS is a kind of spectroscopy, which measures the change in refractive index due to heating of a medium by light. A refractive index, since a gradient exists adjacent to the test sample surface. Hence, a probe laser beam is refracted or bent in a manner proportional to the temperature gradient of the sample near the surface, and this deflection being a measure of the absorbed excitation radiation can be easily determined. It is to be noted that the PDS technique is suitable for studying optically thin samples like RLG mirrors, because of the high sensitivity of the measurements of the absorption. Also, it is suitable for evaluating the samples, through which the transmission cannot be studied.

Photothermal deflection spectroscopy (PDS) is a highly sensitive and precise technique, used to measure the optical absorption and thermal characteristics of a sample. Mostly applications of PDS are based on utilizing a heating beam, which, however, are not accurate and also not suitable for dielectric-coated samples. In such cases, laser cooling of solids, or optical refrigeration as it is called, uses this technique to determine if a laser is cooling or heating a sample. In PDS technique, the laser beam is made to pass near the RLG mirror, and because of the absorption and change in refractive index due to heating, the laser beam is deflected, the deflection determining the absorption. Being a sensitive technique, it can measure small absorption in the coatings. The present conventional PDS methods for laser cooling require multiple laser wavelengths in both the Stokes region and the anti-Stokes region to ensure that cooling is taking place, which can cause problems, if either the lasers must be changed or no lasers are available in the desired wavelength regions.

Fontenot et al. [12] have presented a photothermal deflection technique based on using the deflection of the probe laser to determine if microcooling is occurring inside the sample. Next photothermal deflection spectroscopy (PDS) is one of the most sensitive techniques of the molecular absorption spectroscopy, which is a type of photothermal spectroscopy, that includes photoacoustic and thermal lensing, both very useful and effective. These techniques make use of the absorption of electromagnetic radiation by a medium, leading to a part or the whole of this energy being converted to thermal energy by non-radiative processes. In case of PDS, the source of the electromagnetic radiation used is a modulated laser beam, termed as pump beam. This produces modulated thermal gradient, which is detected by a passing probe beam, mostly a stable HeNe laser, located either inside the transparent media or adjacent to the medium. These thermal gradients result in some changes of the index of refraction, causing the probe beam to be spatially deflected. Finally, the amplitude of this deflection is measured by using a position-sensitive detector, which is used to determine the optical absorption of the sample. PDS technique is of two types: collinear or transverse. In the transverse setup, measurements are done for the opaque solids, by propagating the probe beam through a transparent medium, that is in contact with the surface of the solid, being irradiated by the pump beam at the normal incidence. However, in case of the collinear PDS, the two beams are aligned, so that they are parallel and overlap within the sample. This setup gives better results, since because the beams have a larger interaction distance, thereby producing a larger deflection signal. Presently, PDS technique is mostly used, as it has many advantages including high spectral, spatial, and temporal resolution; high sensitivity; and more importantly, it being a noncontact, and nondestructive technique. Therefore, this technique is applied to measure the optical, thermal, and thermo-elastic materials properties of materials, and thin films. In the photothermal deflection technique, an important point to be noted is that in both collinear and transverse PDS, the surface is heated using a periodically modulated light source, e.g., an optical beam passing through a mechanical chopper or regulated with a function generator. In addition, a lock-in amplifier is used to measure the deflections observed at the modulation frequency. In some experimental setups, a pulsed laser is used as the excitation source, and in that case, the average is taken to measure the temporal deflection of the probe beam to the excitation radiation. In such a setup, the signal falls off exponentially as a function of frequency, and so frequencies around 5–12 Hz are frequently used. Interestingly, PDS analysis of thin films can also be done by using a patterned substrate, supporting optical resonances, such as guided-mode resonance and whispering gallery modes; and coupling the probe beam into a resonant mode. Since the coupling efficiency is observed to be highly sensitive to the incidence angle, the photoheating effect, results in changing of the coupling efficiency, which is characterized to compute the thin film absorption. Photothermal deflection technique is used to measure the local temperature gradients induced by the laser inside the colloid. The experimental setup of the photothermal deflection is shown in Fig. 6.13.

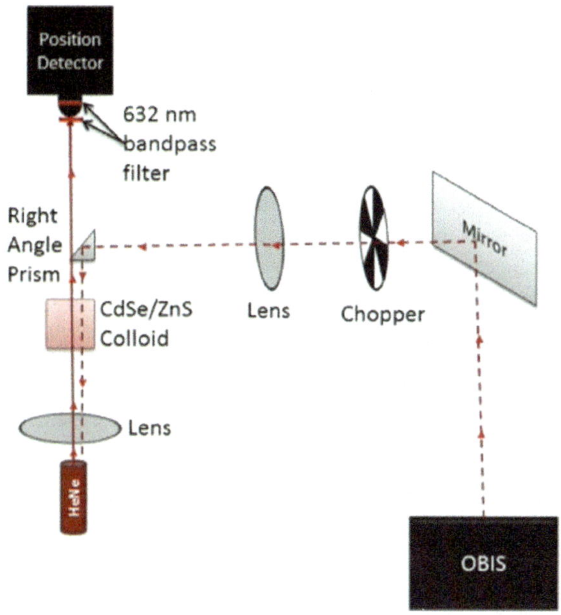

Fig. 6.13 Diagram showing the spectroscopy (PDS) setup. *Figure courtesy* Fontenot et al. [12]

A mode stabilized HeNe laser probe beam is aligned such that the beam passes through the CdSe/ZnS colloid and falls on the center of a position-sensitive detector. The pump beam is co-aligned about 100 μm to the right of the probe beam as it passes through the sample. Care is taken to move it in a counter-propagating position for minimizing the crosstalk with the probe beam in the detector. Also, the designer has to set each laser to its maximum power (e.g., 120 mW for the OBIS 685 nm). The beam is conveniently aligned by moving a right angle prism with a linear actuator. The beams are focused at the edge of the quartz cuvette containing the CdSe/ZnS colloid, by using long focal length lenses. Finally, the angular deflections of the HeNe probe beam, caused by the thermally induced refractive index gradients in the colloid, are measured by using a position-sensitive detector and recorded either by using an oscilloscope or a dual-phase lock-in amplifier of high quality. Also, an optical chopper is set to some suitable frequency (e.g., 10 Hz) to modulate the pump beam to avoid the effect of noise.

6.3.15 *Vibrational Spectroscopy*

Vibrational spectroscopy is considered to be very useful for characterizing thin films of many types. Though optical vibrational techniques are less sensitive as compared to the electron-based spectrometric methods, these are frequently employed for thin-film characterization because of the specific and characteristic vibrational spectra shown by various functional groups and molecules present in the film. Some of the commonly used vibrational spectroscopic techniques are infrared (IR) and Raman spectroscopy. Due to the absorption of IR by the underlying substrate, IR reflection-adsorption spectroscopy (IRRAS) and its polarization modulation (PM) analog, i.e., PM-IRRAS, which uses the polarization selectivity of surface adsorption, are found to be very effective and suitable to characterize thin films.

The IRRAS spectroscopic analysis of thin films is based on using a well-defined p- or s-polarized light, which is directed onto the sample surface at a grazing angle. The working of the technique is simple, in that, the vibrations of molecules from surface layers of the films, which are adsorbed on IR nontransparent planar surfaces, particularly metal surfaces characterized with high reflectivity and are recorded in reflection mode at an angle equal to the angle of incidence. Theoretical interpretation of the spectra is done by measuring the wavelength dependence of reflectance R, which can be defined [13] as:

$$R = I\text{reflected}/I\text{incident} \tag{6.7}$$

where Ireflected and Iincident are, respectively, the intensity of the reflected and the corresponding incident light beams on the sample surface at a grazing angle. Due to the absorption of light by the thin film, a change in reflection occurs, which is given by the following equation:

$$\Delta R/R_0 = (R_0 - R)/R_0 = (4n_{13}/n_{23}).(\sin 2\psi_0/\cos\psi_0)at, \tag{6.8}$$

where R and R_0 are the reflectance with and without the thin film, n_1 and n_2 are the refractive indices of the ambient medium and the film, ψ_0 is the incident angle, a is the absorption coefficient, and t is the film thickness. Hence, the measurement of the change in reflection in the IRRAS analysis provides the direct value of thin-film thickness.

In a PM-IRRAS technique, an IR beam is directed onto the sample surface after passing through a polarizer and a photoelastic modulator (PEM), and after reflection from the sample is collected by an IR lens (BaF2) into a detector. The whole instrumentation setup is evacuated to avoid background atmospheric adsorptions. It is to be noted that, the intensity differences between the s- and p-polarized lights using the PEM and the sum of their intensities are used as reference, and hence, the setup does not require any additional substances for reference.

Schematic of PM-IRRAS setup as been shown below (Fig. 6.14):

$$I(t) = [I_p + I_s + (I_p - I_s)\cos(V_m/\lambda.\cos(\gamma t))]/2, \tag{6.9}$$

where $I(t)$ is the intensity of the output light at time t; I_p and I_s are, respectively, the intensities of p- and s-polarized lights; γ is the modulation frequency; V_m is the modulation voltage; and λ is the wavelength of the IR light.

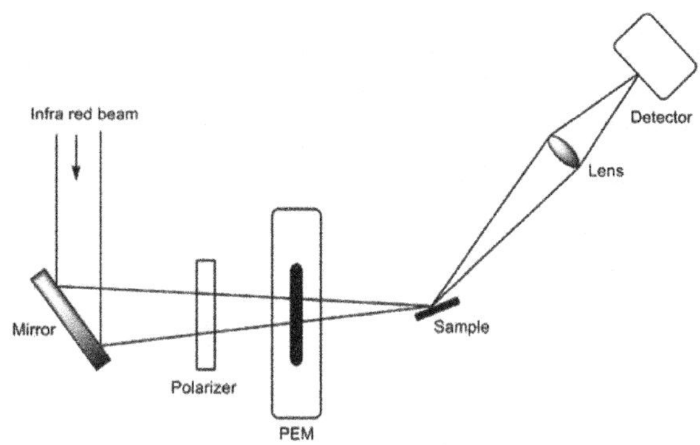

Fig. 6.14 Schematic of PM-IRRAS setup. *Figure courtesy* Schnabel et al. [13]

6.3.16 Ellipsometry

Ellipsometry is an important optical technique used for investigating the dielectric properties, e.g., complex refractive index or dielectric function of thin films. Ellipsometry is based on measuring the change of polarization of polarized light on reflection from the sample or after transmission through it, and then comparing it to some model. It has many uses including characterizing the composition, roughness, thickness, crystallinity, doping concentration, electrical conductivity, and some material properties. This is due to the fact that it is very sensitive to the change in the optical response of incident radiation, which interacts with the material that is being tested. The utility of the instrument can be judged from the fact that a spectroscopic ellipsometer is available in most thin-film laboratories doing analytical studies, which is really very critical for those involved in making RLG mirrors, and other related optics. Ellipsometers are becoming useful to the researchers engaged in other important subjects including biology and medicine, which present many new challenges for the technique, especially in the measurements on the unstable liquid surfaces, and in microscopic imaging, which is immensely useful in medical field.

Basic principle: The basic principle of the device is that it measures the signal in the form of a certain change in polarization, after the incident light in a known state of polarization interacts with the material structure of interest, e.g., reflected, absorbed, scattered, or transmitted. It is possible to quantify the polarization change by the amplitude ratio, ψ, and the phase difference, Δ, explained below.

Since the received signal depends on the thickness of the sample, and also its material properties, ellipsometry has been established as an efficient tool for carrying contact-free determination of thickness and the optical constants of thin films of all types, and more so for RLG coatings. Importantly, the analysis of the change of polarization of light by ellipsometer is able to provide information about the layers, even thinner than the wavelength of the light used for probing and analyzing, even for the single atomic layer. It can also compute the complex refractive index or the dielectric function tensor, giving access to the fundamental physical parameters like mentioned and discussed here. It is used as a matter of routine for characterizing the

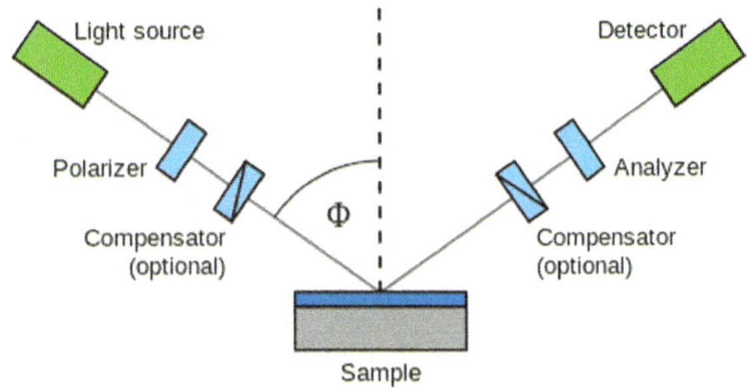

Fig. 6.15 Schematic of an ellipsometry setup for measuring the optical constants of thin films. *Figure courtesy* Ellipsometry wikipedia.org

film thickness for single layers or multilayer stacks (as in RLG mirrors) in a broad range from a few angstroms to a few micrometers with great accuracy.

Experimental details: In the laboratories, ellipsometry is mostly studied in the reflection setup. The exact change in the nature and degree of the polarization change depends on the thickness, complex refractive index or dielectric function tensor of the sample. It is important to note here that though mostly optical techniques are diffraction limited, ellipsometry is unique in the sense that it exploits the phase information in the form of the polarization state and can provide sub-nanometer resolution. Thus, we see that the technique is useful for investigating thin films thickness of less than a nanometer to a few micrometers. The technique is based on the model of assuming that the sample is composed of a small number of optically homogeneous and isotropic discrete, well-defined layers. The experimental setup is shown in Fig. 6.15.

The electromagnetic radiation of suitable wavelength is emitted by a light source and is linearly polarized by a polarizer. Then, it is made to pass through a compensator, which is chosen as a retarder or a quarter wave plate, and is incident onto the sample. After reflection, the radiation is made to pass through a compensator and a second polarizer, called an analyzer (polarizer and analyzer perform different operations, but are otherwise identical) and is finally incident on the detector. In some designs, the compensators are replaced by a phase modulator in the path of the incident light beam. It is to be noted that the ellipsometry is a specular optical technique, which means that the angle of incidence is equal to the angle of reflection. In fact, the plane containing the incident and the reflected beams is termed as the plane of incidence. The polarized light is of two types: Light polarized parallel to this plane is called p-polarized, and the one polarized in a direction perpendicular to this plane is called s-polarized.

Data acquisition: The next step is to measure the complex reflectance ratio of a system, which is given in terms of the amplitude component and the phase difference. It is to be noted that the polarization state of the light incident on the sample may be decomposed into s-polarized and p-polarized components, the s-polarized component oscillating perpendicular to the plane of incidence, and the p-polarized component oscillating parallel to the plane of incidence. The amplitudes of the s- and p-components, after reflection and normalization to their initial value, are denoted by r_s and r_p, respectively. The angle of incidence is chosen close to Brewster's angle of the sample to obtain a maximal value of the difference in r_s and r_p. Ellipsometer measures the complex reflectance ratio which is the ratio of r_s and r_p, and the phase shift. It is important to note here that since ellipsometer measures the ratio, or the difference of the two values, and not their absolute values, it gives very accurate and reproducible results. For instance, since it is quite insensitive to scattering and fluctuations, it requires no standard sample or reference beam.

Data analysis: Since ellipsometry is an indirect method, implying that measured r_s and r_p values cannot be converted directly unto optical constants of the RLG mirror or any other sample; and hence, a model analysis must be performed, which is a difficult exercise and requires computer software. However, direct conversion of r_s and r_p values is possible in very simple cases of isotropic, homogeneous, and infinitely thick films. In other cases, some layer model, and Fresnel equations have to be used to find the optical constants, which are matched and verified by experimentally achieved values. This exercise can be performed only by a skilled and experienced optical engineer.

Some companies especially M/s Woollam are manufacturers of very efficient spectroscopic ellipsometers for non-destructive thin film and bulk material characterization. Spectroscopic ellipsometry (SE) has become the standard technique for measuring thin film thickness and optical constants (n and k). Spectroscopic ellipsometry is being used for the characterization of all types of materials including dielectrics, semiconductors, and metals. M/s Woollam offer ellipsometers in a broad spectral range from the UV to the far IR. In addition, some very sophisticated models are available, which are suitable for accurate work, as in case of RLG mirrors, which are provided with automated sample translation, variable temperature, and focusing optics. As discussed above, ellipsometry mainly aims at measuring as to how the p- and s-polarized components change upon reflection or transmission w.r.t. to each other, and therefore, the reference beam is an important part of the experiment. The technique is simple in that a known polarized light is reflected or transmitted from the sample and the output polarization is measured. The change in polarization is measured by the ellipsometer and given by:

$$\rho = \tan(\psi)e^{i\Delta}, \tag{6.10}$$

A sample ellipsometry measurement setup is shown in Fig. 6.16. The incident light is linearly polarized with both p- and s-components. Upon reflection, the light undergoes amplitude and phase changes for both p- and s-polarized components, which are measured by the ellipsometer. The schematic of a typical ellipsometry configuration is shown below.

As discussed above, linearly polarized light is mostly reflected from the sample surface, and the polarization change is measured to determine the sample response. The main tools in this ellipsometer for collecting ellipsometry data include: the light source, polarization generator (polarizer), sample, polarization tester (analyzer), and detector. The polarizing and analyzing optical components for manipulating the polarization include polarizers, compensators, and phase modulators. In some

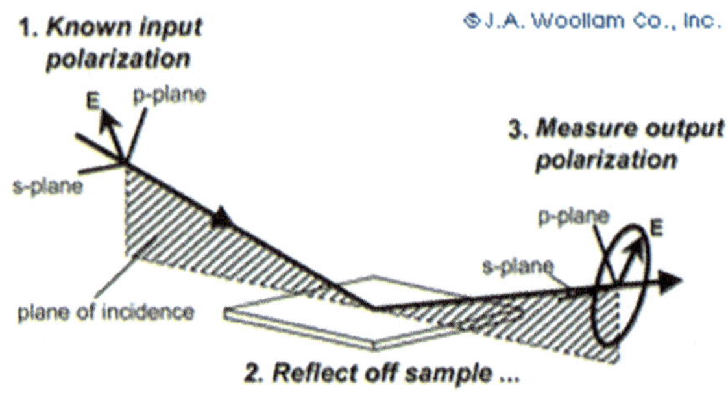

Fig. 6.16 Commonly used ellipsometry configuration. *Figure courtesy* M/s Woollam

ellipsometer configurations, design is made of including rotating analyzer (RAE), rotating polarizer (RPE), rotating compensator (RCE), and phase modulation (PME). The schematic of such an ellipsometer setup is shown below:

This configuration uses a polarizer to define the incoming polarization and a rotating polarizer after the sample to analyze the outgoing polarization. The detector converts light into a voltage, the degree of whose change gives the measurement of the reflected polarization. The RAE configuration is shown in Fig. 6.17, in which a suitable light source produces unpolarized light which is then made to pass through a polarizer to get polarized light, since the polarizer allows light of only a preferred electric field orientation to pass. The polarizer axis is oriented between the p- and s-planes in such a way, that both arrive at the sample surface. The linearly polarized light after reflection from the sample surface becomes elliptically polarized and passes through a continuously rotating polarizer, called the analyzer. The amount of light allowed to pass depends on the orientation of polarizer relative to the electric field ellipse received from the sample. The function of the detector is to convert light into electronic signal to find the reflected polarization. Finally, this information is compared with the known input polarization to determine the change in polarization resulting from the sample reflection, which is the ellipsometer measurement of ψ and Δ.

In situ applications: The application of in situ spectroscopic ellipsometry has also become quite popular in the sophisticated laboratories, to process monitoring and control the multilayer dielectric coatings as in case of RLG mirrors. The underlying advantage in this case is that after the completion of the deposition of each layer, depending upon the feedback, corrective action can be taken for the deposition of each layer. The in-situ SE can simplify the routine process of characterization and

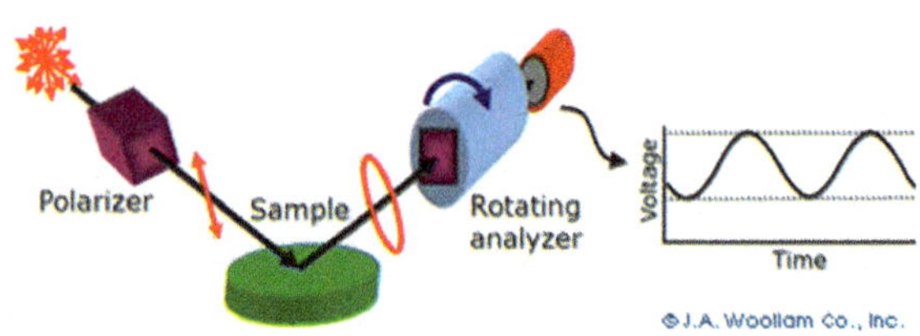

Fig. 6.17 Schematic of rotating analyzer ellipsometer. *Figure courtesy* M/s Woollam

development. Recently, many innovations have been incorporated into the software available with the equipment. The virtual interface (VI) models are available commercially, to readily determine deposition rates of each layer and optical constants of the topmost film without requiring a detailed model of underlying layers. These VI models can perform sequential analysis of each layer of a multifilm stack while they are deposited during a single run. Hence, the designer can vary the process control parameters for each layer during the deposition and thus can easily characterize film deposition rates and optical constants for various conditions.

A commercially available instrument [14] provides high-end and reliable state-of-the-art technology for characterizing thin dielectric thin films especially RLG Mirrors. This device combines nulling ellipsometry and microscopy: imaging ellipsometry (IE) to overcome the limits of classical ellipsometry. When linearly polarized light with an axis pointing anywhere other than the s- or p-direction, is incident on a sample, the reflected light generally shows an elliptical state of polarization. Conversely, if the same elliptical state of polarization, but with a reversed sense of rotation, is incident on a surface, it generates a linearly polarized light on reflection. This is the principle of nulling ellipsometry.

Imaging ellipsometry is a technique, which combines spatial evolution of optical microscopy with thin-film measurement characteristics of ellipsometry. The technique has been used with a lot of interest for a wide range of applications including biotechnology and semiconductor metrology. Mainly, IE is used either as a qualitative technique with ellipsometry images showing surface topology variations, or as a quasi-quantitative technique, in which the change in ellipsometry images is used to derive the change in film thickness.

Recently, Negara et al. [15] proposed a new design of an imaging return-path ellipsometer, which enables reflection-based measurements of both flat and curved surfaces. It is known that the return-path ellipsometry (RPE) based on mirrors cannot be applied to the surface characterization of a specimen with a free-form shape. Negara et al. [15] have circumvented this restriction by using microsphere-based retroreflectors instead of a mirror.

6.4 Concluding Remarks

We observe that very serious efforts are being made for evolving more novel techniques of thin-films characterization, both by the researchers and the commercial firms. The quest for improving the performance of the optoelectronic gyroscopes is leading to making serious efforts for fabricating better RLG and FOG components, for which new and improved techniques for optical testing and thin film characterization are being suggested and tried with good results. So, these techniques are complementing each other for providing products with more and more improved quality and performance. Hence, the designers and engineers of the RLGs and FOGs have to be well-versed with all these technologies for finding the ways to develop state-of-the-art products. The various parameters of optoelectronic gyroscopes like scale factor, random walk, and operating life can be improved only if improved optical components and coatings, and also, the optical testing and thin-film characterization techniques are available. For example, we need better mirrors for improving the optical gyroscopes, which require better optical substrates, and better techniques for optical testing, thin films deposition, and thin films characterization. This is the reason that all these technologies are interrelated and hence are rapidly evolving.

References

1. K.N. Chopra, A.K. Maini, *Thin Films and their Applications in Military and Civil Sectors* (Defence Science and Documentation Centre, Metcalfe House, Defence Research and Development Organisation, Ministry of Defence, Delhi, India, 2010)
2. L. Cromer Christopher, E. Hurst Katherine, X. Li, J.H. Lehman, Black optical coating for high-power laser measurements from carbon nanotubes and silicate. Opt. Lett. **34**, 193–5 (2009)
3. A. Ulman, *Characterization of Organic Thin Films* (Momentum Momentum Press, New York, 2010), 296 p
4. C.H. Ahn, K.M. Rabe, J.-M. Triscone, Ferroelectricity at the nanoscale: local polarization in oxide thin films and heterostructures. Science **303**(5657), 488–491 (2004)
5. D. Grosso, A.R. Balkenende, P.A. Albouy, A. Ayral, H. Amenitsch, F. Babonneau, Two-dimensional hexagonal mesoporous silica thin films prepared from block copolymers: detailed characterization and formation mechanism. Chem. Mater. **13**, 1848–1856 (2001)
6. A. Salleo, R.J. Kline, D.M. DeLongchamp, M.L. Chabinyc, Microstructural characterization and charge transport in thin films of conjugated polymers. Adv. Mater. **22**, 3812–3838 (2010)
7. A. von Finck, M. Hauptvogel, A. Duparré, Instrument for close-to-process light scatter measurements of thin film coatings and substrates. Appl. Opt. **50**, C321–C328 (2011)
8. S. Schröder, D. Unglaub, A. von Finck, M. Hauptvogel, M. Trost, T. Herffurth, A. Duparré, J.C. Stover, Angle and wavelength resolved light scattering measurement of optical surfaces and thin films, in *Proceedings of SPIE 8838, Optical Manufacturing and Testing X, 883802*, 7 Sept 2013
9. C. Mühlig, S. Bublitz, W. Paa, Enhanced laser-induced deflection measurements for low absorbing highly reflecting mirrors. Appl. Opt. **53**, A16–A20 (2014)
10. A.J. de Lange, *Spectrophotometer Dynamic Range and Linearity in Brewing Materials and Processes* (2016)
11. S.M. Nilapwar, M. Nardelli, H.V. Westerhoff, M. Verma, in *Methods in Enzymology*, ed. by D. Jameson, M. Verma, H.V. Westerhoff (2011)
12. R.S. Fontenot, V. Mathur, J. Barkyoumb, New photothermal deflection technique to discriminate between heating and cooling. J. Quant. Spectrosc. Radiat. Transfer **204**, 1–6 (2018). https://doi.org/10.1016/j.jqsrt.2017.08.019license
13. T. Schnabel, M. Musso, G. Tondi, Univariate and multivariate analysis of tannin-impregnated wood species using vibrational spectroscopy. Appl. Spectrosc. **68**(4), 488–494 (2014). https://doi.org/10.1366/13-07181
14. Accurion GmbH, *Measure Ultra Thin Films*. www.accurion.com/
15. C. Negara, T. Längle, J. Beyerer, Imaging ellipsometry for curved surfaces. J. Vacuum Sci. Technol. B **ICSE2019**, 014016 (2020). https://doi.org/10.1116/1.5129654@jvb.2020.ICSE2019.issue-1

Appendix A
Technical Terms in the Book

1. Ring laser gyroscope (RLG)

A ring laser gyroscope (RLG) consists of a ring laser having two counter-propagating modes over the same path and is used to detect and measure rotation, the operation of which is based on the principle of the Sagnac effect that shifts the nulls of the internal standing wave pattern in response to the angular rotation. The shifts in the standing wave are shown by the interference between the counter-propagating beams, observed externally.

2. Alignment of Optical Elements in RLG

Alignment of the optical elements is critical for the successful and optimized performance of RLG and must be done very carefully, since the HeNe laser works only when the alignment is very precise. For giving some idea, this process is briefly discussed here. This requires great skill on the part of the experimentalist. Though some instructions for getting a good alignment are known, still a lot of patience for following the trial-and-error technique is required. It is absolutely important that the experimentalist starts the next step only after completing the ongoing step.

3. Frequency synchronization—lock-in—of the two modes

The frequency synchronization—lock-in—of the two modes, encountered at low rates of rotation, is one of the serious problems of ring laser gyroscope designing. The problem of lock-in is due to the weak mutual coupling caused by backscattering in the pressure ~ milli Torr, which results in a dead band around zero, and also a strongly nonlinear signal near the threshold. The mode locking is due to the fact that the backscattering in the optical path (mirrors) weakly couples the counter-propagating beams, and thus in case, the beat frequency gets smaller than a threshold value, and the modes oscillate at the same frequency, eliminating the beat note.

4. Noise due to Optical Kerr Effect

The optical Kerr effect is very important source of noise, since the electric fields of the counter-propagating beams cause changes in the index of refraction, which is

K. N. Chopra, *Optoelectronic Gyroscopes*, Progress in Optical Science and Photonics 11,
https://doi.org/10.1007/978-981-15-8380-3

non-reciprocal if $|E1|$ and $|E2|$ are not equal; e.g., I1/I2 = $\pm 10^{-4}$ leads to 10^{-3} deg/h shifts.

5. Effect of Null Shift and Mode Locking on the Performance of the RLGs

There are many types of sources of error including null shift and mode locking. The null shift arises due to the fact that the beat frequency is nonzero even when the rotation rate Ω=0.

6. Superluminal ring laser gyroscope (RLG)

A superluminal ring laser gyroscope (RLG) by using multilayer optical coatings with huge group delay (GD), p which can measure the absolute rotation with a giant sensitivity-enhancement factor of ~103; while, the broadband FWHM of the enhancement factor can be 20 MHz.

7. A fiber-optic gyroscope (FOG)

A fiber-optic gyroscope (FOG) is an optical device for sensing the changes in orientation, thereby performing the function of a mechanical gyroscope, for its operation is based on the interference of light having passed through a coil of optical fiber of very large length ~5 km.

8. Scale Factor

The scale factor of the FOG K is given by:

$$K = \frac{(4A)}{\lambda P},$$

where A is the area bounded by the FOG and P is the perimeter of the FOG, given by: $P = \pi D$, D being the diameter of the loop, and $AN$is the total enclosed area. The interesting step taken is that the effective area is increased by winding N loops of the fiber-optic cable.

$$\Omega = \frac{\lambda c}{2LD},$$

$L = N\pi D$where L is the total length of the fiber, λ is the wavelength of light in the medium, Ω is the angular rate of rotation,, and $\Delta\phi$ is the phase difference introduced, which is given by:

$$\Delta\phi = \frac{2\pi L D \Omega}{\lambda c}$$

9. Figure of merit (FOM)

Figure of Merit is an important consideration for the choice of the substrate material, and is given by:

$$F = \frac{K}{B_e \alpha (1 + \nu)},$$

where K = thermal conductivity, B_e = effective absorption, α = linear expansion coefficient, and ν = Poisson's ratio.

10. Double clad fiber

Double clad fiber is an important class of optical fibers, which has a core doped with the rare earth. In case of standard single-mode fibers, pumps needed to be used as pump sources do not have enough power to achieve the desired output power levels. Double clad fibers are able to overcome this problem by having a doped single-mode core with two layers of cladding around the core. The advantage is that the inner cladding can accept large amounts of pump light from multiple pump source coupled into the inner cladding layer. Thus, the double clad approach provides optimization of high power and high beam quality.

11. Mode instabilities in a Fiber

Above a threshold power level that depends on a number of parameters, instabilities can cause a good single-mode beam to degrade into a poor-quality, unstable multi-mode signal. This transverse mode instability has been understood as the thermal effects occurring after power density in the fiber exceeds a threshold value, below which the laser generates a stable single-mode beam of high quality.

12. Optical Nondestructive Testing (NDT)

Optical nondestructive testing (NDT) has drawn the attention of many workers in recent years because of its nondestructive imaging characteristics with high precision and sensitivity. There are many types of NDT technologies like fiber optics, electronic speckle, infrared thermography, endoscopic, and terahertz technology. Each technique has its own advantages.

13. Optimum Pair Design for Dielectric Mirrors

Designing of a multilayer stack for the laser mirrors is in the form of a quarter wave stack consisting of the layer sequences as given below:

$$[\text{Air}][HL]^n\text{Glass}$$

where Hdenotes a high index layer of quarter wave optical thickness (QWOT), Ldenotes a low index layer of quarter wave optical thickness (QWOT), and $[HL]$denotes the basic period, which is repeated n times.

14. Laser Speckle Interferometry and Speckle NDT

Laser speckle interferometry is based on using the intensity distribution pattern formed from space interference, generated by illumination of coherent light onto rough object surfaces, and is considered as an effective NDT technology with many advantages including non-contact, no requirement of the illumination over the surface of the object to be uniform, high sensitivity, and detection rate, and as a result it has been widely used in various industries.

15. Electronic Speckle Pattern Interferometry (ESPI) Technique

The electronic speckle pattern interferometry (ESPI) technique is used for measuring changes in diffuse surfaces. The working of a dynamic phase-shifting electronic speckle pattern interferometer is based on producing of a 9 ns pulse during each measurement and then capturing four phase-shifted interferograms during the single pulse.

16. Optical Time-Domain Reflectometer (OTDR)

An optical time-domain reflectometer (OTDR) is a single-ended measurement tool, which provides a detailed picture of section-by-section loss and operates by sending a high-power pulse of light down the fiber and measuring the light reflected back, thereby determining the distance (between launching point and the event) of each event by using the time taken for individual reflections to return. In this technique, the time corresponds to the acquisition duration, i.e., the period of averaging the results.

17. X-ray Photoelectron Spectroscopy (XPS):

Photoelectron spectroscopy is based on utilizing photoionization and analysis of the kinetic energy distribution of the emitted photoelectrons, thereby evaluating the composition and electronic state of the surface region of a sample.

18. Reflection High-energy Electron Diffraction (RHEED)

RHEED is a technique useful for characterizing the surface of materials. RHEED systems gather information only from the surface layer of the sample, which separates it from other materials characterization techniques that also rely on the diffraction of high-energy electrons.

19. Atomic Force Microscopy (AFM)

Atomic force microscopy (AFM) also known as scanning force microscopy (SFM) is a very high-resolution type of scanning probe microscopy, with resolution ~ fractions of a nanometer, i.e., more than 1000 times better than the optical diffraction limit. The AFM is a very important tool for imaging, measuring, and manipulating thin films at the nanoscale level.

20. Fourier Transform Infrared (FTIR) Technique

Fourier transform infrared (FTIR) spectroscopy is a technique which is used to obtain an infrared spectrum of absorption, emission, photoconductivity, or Raman scattering of a thin film. Fourier transform infrared spectroscopy is a combination of IR absorption spectroscopy and a Michelson interferometer, based on recording the absorption as a function of time, according to the movement of the mirror in one of the interferometer arms.

21. Differential Interference Contrast (DIC) Microscopy

Differential interference contrast microscopy, also known as Nomarski interference contrast (NIC) microscopy or simply as Nomarski microscopy, is indeed an important optical microscopy illumination technique used for enhancing the contrast in unstained, transparent samples. DIC is based on the principle of interferometry to gain information about the optical path length of the sample and thus to see features, which are otherwise invisible.

22. Spectrophotometers

A spectrophotometer is an important instrument for characterizing the RLG mirrors, which has a monochromator, for producing a light beam having wavelengths in a narrow band around the desired selected wavelength, and a component for measuring the ratio of that beam's intensity as it enters and leaves a cuvette.

23. Cavity ring-down spectroscopy (CRDS)

CRDS provides an ideal solution for determining the reflectance of highly reflective optics by measuring the total loss of the optic, which includes absorption, transmission, and scatter, from which the reflectance of the optic can be determined by subtracting the measured loss from 100%.

24. Photothermal spectroscopy

Photothermal spectroscopy is a group of various high-sensitivity spectroscopy techniques, followed in laboratories for evaluating optical absorption and thermal characteristics of a sample, e.g., a multilayer coated sample like RLG mirror. These techniques are based on the change in the thermal state of the sample due to the absorption of radiation.

25. Photodefletion Spectroscopy (PDS)

In PDS technique, the laser beam is made to pass near the RLG mirror, and because of the absorption and change in refractive index due to heating, the laser beam is deflected, the deflection determining the absorption.

26. Vibrational Spectroscopy

Vibrational spectroscopy is considered to be very useful for characterizing thin films of many types. Though optical vibrational techniques are less sensitive as compared to the electron-based spectrometric methods, these are frequently employed for thin-film characterization because of the specific and characteristic vibrational spectra shown by various functional groups and molecules present in the film. Some of the commonly used vibrational spectroscopic techniques are infrared (IR) and Raman spectroscopy. Due to the absorption of IR by the underlying substrate, IR reflection–adsorption spectroscopy (IRRAS) and its polarization modulation (PM) analog, i.e., PM-IRRAS, which uses the polarization selectivity of surface adsorption, are found to be very effective and suitable to characterize thin films.

27. Ellipsometry

Ellipsometry is an important optical technique used for investigating the dielectric properties, e.g., complex refractive index or dielectric function of thin films. Ellipsometry is based on measuring the change of polarization of polarized light on reflection from the sample or after transmission through it, and then comparing it to some model. Imaging ellipsometry is a technique, which combines spatial evolution of optical microscopy with thin-film measurement characteristics of ellipsometry. The technique has been used with a lot of interest for a wide range of applications including biotechnology and semiconductor metrology. Mainly, IE is used either as a qualitative technique with ellipsometry images showing surface topology variations, or as a quasi-quantitative technique, in which the change in ellipsometry images is used to derive the change in film thickness.

Appendix B
Firms Engaged in Working on RLGs and FOGs

There are many firms engaged in working on ring laser gyroscopes and fiber-optic gyroscopes. The notable firms are listed below:

- Ring Laser Gyroscope Firms:

1. Heppell Photonics GmbH
2. Ericco International Limited
3. Honeywell International Inc.
4. Teledyne CDL, Inc.
5. Mitsubishi Precision Co., Ltd
6. Safran Electronics and Defense SAS
7. Northrop Grumman Corporation
8. Optics Blazer AG
9. Kearfott Corporation
10. Sperry Marine B.V.
11. Xsens Technologies B.V.
12. Memsense, LLC.
13. Rieker, Inc.
14. Jewell Instruments, LLC.
15. Cooner Wire Company
16. Sensata Technologies
17. ACE Controls Inc.
18. RS Components, Ltd.
19. Zurich Instruments

    - Fiber-Optic Gyroscope Firms:

1. M/s DK Photonics
2. Clear Align—Eagleville, PA. • Established in 2003.
3. General Photonics Corp., Div.
4. KVH Industries Inc.
5. Optics Valley Metering and Calibration Co.

K. N. Chopra, *Optoelectronic Gyroscopes*, Progress in Optical Science and Photonics 11,
https://doi.org/10.1007/978-981-15-8380-3

6. Optolink Ltd., Optical Component—Washington, England.
7. Rayzer Optical Technology Co.
8. Silicon Valley Photonics Ltd.

The manufacturer's authorised representative in the EU is Springer
Nature Customer Service Centre GmbH, Europaplatz 3, 69115 Heidelberg,
Germany. If you have any concerns regarding our products, please
contact ProductSafety@springernature.com

Printed and bound by CPI Group (UK) Ltd, Croydon, CR0 4YY
30/09/2026
02225265-0009